ROBOTICS

Programming and Math

A Guide for Teachers and Students

By Robin Sitzler-Frazier M. Ed

Robotics Programming and Math
A guide for Teachers and Students

ISBN-13: 978-1542398695
ISBN-10: 154239869X

Library of Congress Control Number: 2017900414

ACKNOWLEDGMENTS

I would like to thank Mr. Christopher Fultz for his contribution to Chapter 4 "Teamwork" addition to this book. Mr. Fultz is a Robotics Competition Cyber Blue 234 Team Mentor at Perry Meridian High School in Southside Indianapolis, Indiana. He is currently leads FIRST® Team students in First LEGO League Competitions, and professionally; he is Head of Program Management, RR Defense Systems. The importance of his "White Paper" helps to teach students that Continuous Improvement is a learning tool for improving your mistakes.

"Cyber Blue is focused on exceeding known limits. We concentrate on education, community, and competition. Whether we are inviting the community into our lab to experience robotics firsthand or helping out our neighboring teams in *FIRST*®, Cyber Blue is determined to be a force and recognized name in our Southside Indianapolis community". http://cyberblue234.com/author/cfultz/

I would like to also thank Plasma Robotic of Mesa Arizona for the cover picture of their 2011 Robotics Team event. FLL-Tournament, Saturday, December 7, 2013 9:35 AM

"Plasma Robotics is a FIRST Robotics and Zero Robotics Team from Mesa, Arizona. Students from Red Mountain High School and other nearby schools such as Mountain View High School and the Mesa Academy involve themselves in the widespread robotics program, learning the engineering process and having fun while doing it! In our 10th year, Plasma hopes to continue to have a positive impact on its team members and the community." http://www.plasmarobotics.com/

TABLE OF CONTENTS

CHAPTER 1

What are Robots?

Have you ever owned a robot and did not realize it was a robot? How about a remote controlled car? A robot is a man-made machine that can perform work or other actions usually performed by humans, either automatically, autonomous, or by remote control. A robot is a programmable device built with mechanical features that can perform tasks and interact with its environment. The definition includes "without the aid of human interaction," however; we need humans to build and program the robots. The word robot comes from the Czech word "robota" meaning "forced work or labor" or "compulsive servitude." The playwright Karel Capek (1921) coined the term. The play, "Rossum's Universal Robots" is an unusual story about manufactured human-like servants and their struggle for freedom.[1]

The science and study of robots are referred to as robotics. We need robots because they can work more economically than we can, because it is easier for a robot to work more cautiously than a human. Have you ever heard someone complaining that his or her feet hurt from standing all day? A robot will not ever have pain or complain. It might need to be oiled, or parts exchanged for newer ones, but the human body will wear out much faster than any robot. Robots can also get into tight places, travel to outer space, and fly into unexplored territories, such as a glacier, or volcano.

Do you dread having your parents tell you to do the same thing over and over again? Well, we can program a robot to do the same things repeatedly, by adding a loop block to the program. A repeat code in your program gives robots the instructions to perform functions for long hours without intervention. They do not get sick or need a vacation. We need them for these reasons, just as much as we need trained programmers, coders, and engineers, to design, build and program them.

---

[1] "The Curious Origin of the Word 'Robot'." Interesting Literature. N.p., 21 Sept. 2016. Web. 03 Aug. 2016.

The Intelligent Robot

The controller is a main feature that makes robots fun to build and easy to use. A remote controller, infrared receiver, or what LEGO Mindstorms calls the 'Intelligent Brick Brain," are the parts of the entire robot, in which you can build and drive almost instantly. You may already know how to play video games. The remote controller that pairs with a brick brain is similar, yet with more features that control robots and the custom controller designed ergonomically for handheld use. A remote controller may have two analog joysticks, which provide exceptional control for accurately driving and turning a robot, or raising a robotic arm with precise control.

("The Intelligent Robot", 2016).[2]

Extra buttons built into the controller provide additional controls for closing a claw or activating a lifting mechanism with varying degrees of freedom (DOF). Just as a human can move in various degrees of freedom with rotating joints, a robots built with movable parts have the ability to move in single independent directions of motion. To be able to move in various directions means something has many DOF. Up, down, or left and right, are DOFs.

---

[2] Creative Commons License: http://www.rapidonline.com/catalogueimages/product/S70-6275P01WL.jpg

Robotic Arm- Degrees of Freedom

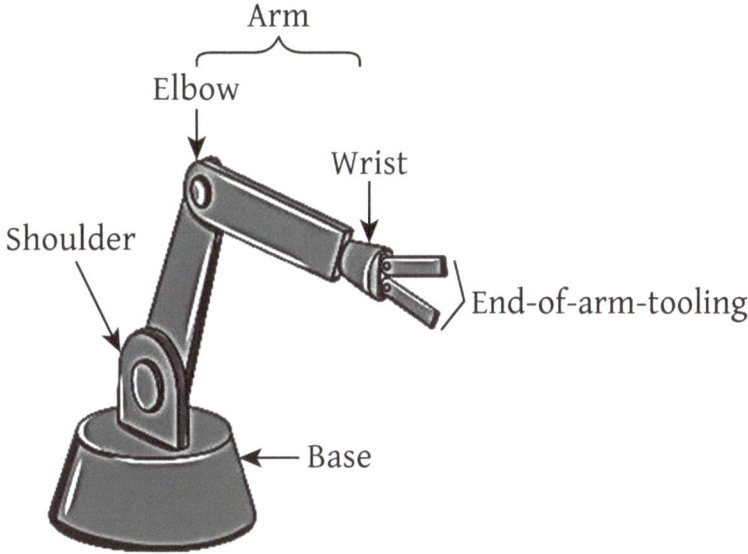

What Is a Controller?

A robot controller is a general category that combines hardware and software to program single and multiple (integrated) robots. One example is an industrial robot controller. It does the repetitive tasks designed from a central processing unit (CPU) board, (the part of the computer micro-processing chip that does most of the data processing) and is attached to the servo motor, and direct current (DC) hubs. In a more defined explanation, a robot controller is in charge of controlling servo motors. Robots are built with different types of servomotors called **rotary actuators** or **linear actuators**. They allow you to control the angular and linear positions, velocity, and acceleration with precise control. You can combine a suitable motor with a sensor to position your mobile robot, or get feedback.

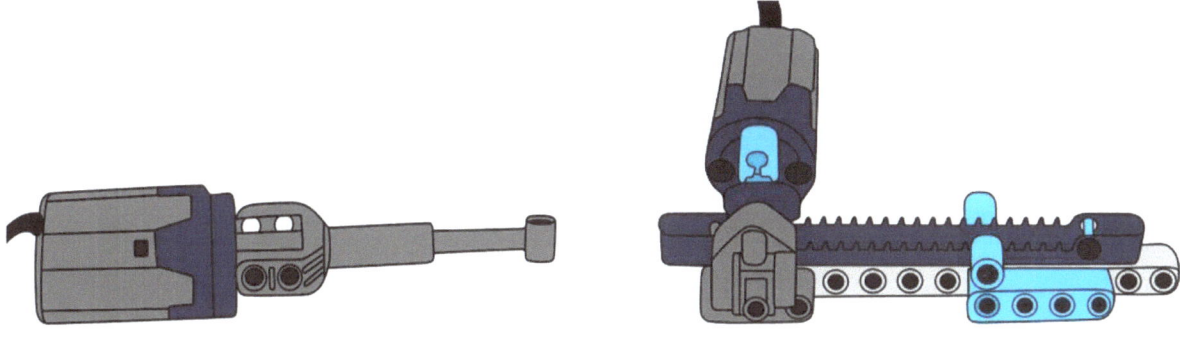

Lego Simple Linear Actuator   *Rack and Pinion Simple Actuator*

Programmable Autonomous Robots

The packaging of golf balls is one way of integrating **autonomous** (independent once it is programmed) robots in manufacturing. **B & R Automation** is an international company that builds automated robotic systems for industries. The company has designed three robots to work precisely together—*an articulated robot arm,* a *tripod robot*, and a *Cartesian coordinate robot* complete with kinematic characteristics, which are synchronized to pick up and package golf balls. The program is written with an industrial PC and designed with speed and efficiency that quickly packages the golf balls for retail sales. Below is a video link showing how it works.[3]

The Vex IQ® is a wireless robot that uses a remote controller that pairs with the brick brain. The brain is a Texas Instruments MSP430 microcontroller that reads the user's inputs and transmits them wirelessly, or through the Tether port. It does all this while using a built-in 900 MHz radio. A remote Autopilot robot with or without sensors can affect the intelligence of the robot.

Picture of the VEX IQ Brain

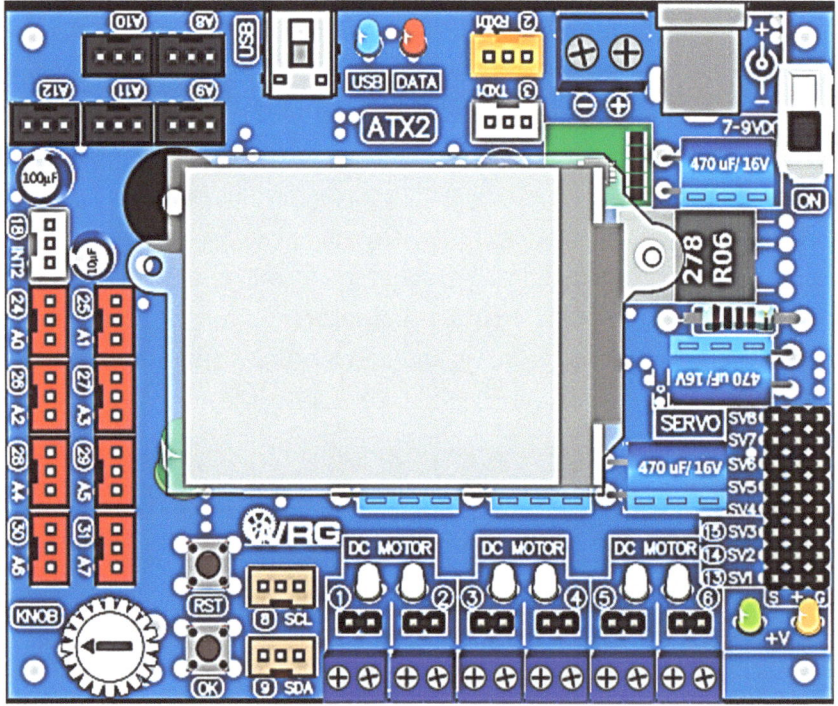

("Picture of the VEX IQ brain", 2016).

---

[3] https://www.youtube.com/watch?v=0_2qkyL3GOk

Teleoperation

The term **tele-operation** was coined by Edwin G Johnsen in 1984. Teleoperation refers to a machine that can be operated at a distance. You may know about remote controls and have some robots that use remote controls. Tele-operation is a phrase that is used in academic research or technical environments. It is used in reference to automation machines or robots. The history of radio communications can be traced back to Nikola Tesla; he developed some of the basic principles and systems of teleoperations during the late 1800s ("Radio", 2016).[4]

The area of **tele-robotics** is concerned with semi-autonomous control from a distance, chiefly using wireless networks. Different types of wireless networks include Bluetooth, the Deep Space Network, and other similar tethered connections. There are two major subfields, tele-operation and tele-presence; however, we will use the word controller in place of tele-operation mostly, and discuss in more detail the types of robots and how to control them in the next chapters.

What is a Microcontroller?

A microcontroller is the "brain" and control center of a robot. The brain is built to be responsible for all calculations, decision making, and communications that execute the program (i.e. sequence of instructions) coded into its brain. How it is designed mechanically is that the microcontroller has a series of pins of electrical signal connections that can be written as code.

Algorithms

The decision-making and communications that execute the sequence of instructions to the brick brain first and secondly into the sensor controls is known as an **algorithm**. The rules are precise and specify how to perform tasks. Algorithms are the core of computer science. Information can be programmed (*coded*) by telling the computer or microcontroller what steps to perform. In ROBOTC, the algorithm can be seen as ordinary sentences in any human language that works as a flow chart. The "**pseudo-code**" portion of the algorithm is like an outline of the program, written from spoken language (our regular language) with common words. It will be converted into a real programming statement. More details on programming later.

Technology advances have allowed microcontrollers to be programmed with various high-level languages including C, C++, Python, Java, Net and Basic. You should be able to find a programming language that works for your robot needs and programming experience. For the novice robot builder, several easy graphical programs can work with your robot. Some examples are:

> **RobotC-** is one of the most recognized programming languages used in educational robotics and competitions. It is a C-Based Programming Language with a simple software drag and drop environment and easy to learn.

---

[4] Inventing the Temper-O-Meter, an IoT Marvel | Viget. (2016). https://viget.com. Retrieved 29 December 2016, from https://www.viget.com/articles/inventing-the-temper-o-meter-an-iot-marvel

SCRATCH- ("Scratch - Imagine, Program, Share", 2016) works with LEGO motorized models. You can also program interactive animations using stories and games. Additionally, with Scratch, you have the chance to share your creations with other people in the online world.

Ardublock- blocked-based graphical programming environment for creating Arduino programs, it allows the user to see the Arduino code and download the program.

MBlock- graphical programming environment that is based on Scratch 2.0, the interface is well set up and easy to use by everyone. mBlock is compatible with many software programs; it supports Arduino Leonardo, Arduino Uno boards, Arduino Mega128-, Arduino Nano, Arduino Mega 2560, PicoBoard, mCore (Based on Arduino Uno) which work with LEGO NXT and Vex IQ models.

Why choose microcontrollers over computers alone? In K-12 education, microcontrollers are being used to teach Science, Technology, Engineering, Math (STEM) and STEA (ART) M initiatives and can easily integrate into school technology classrooms. Here are some other advantages:

Microcontrollers can demonstrate the essentials—variables, constants, operators, counters, open and closed loops, conditional branching, and subroutines or function calls—for programming.

Microcontrollers interact with circuits through their input/output pins for controlling lights, pushbuttons, beepers, sensors, and motors. We will learn more about these in the next chapters.

Microcontrollers are everywhere—in appliances, toys, and personal electronics. Learning to program them will give you a hands-on approach to understanding how things work.

Microcontrollers are used in equipment employed in many career fields — they can be built for use in agriculture, medicine, health sciences, media construction, environmental studies, engineering, forensics, manufacturing, media, music, scientific research, and other industries.

Teamwork

Sometimes students come to a robotics class with high expectations and they can be very independent thinkers. They soon find out that they must learn to work together as a team, ask for help, and offer to help others. Team building is essential to a school class, or a group formed to learn about robotics. It takes a team to build robots. Some students do not know much but can figure out the math, while others have an enormous amount of experience using technic bricks to build and create the robot. The team shares ideas, and how valuable it can be to learn from the mistakes of others. Why? There is a benefit to working with others on something that you do not have much experience. Teams can quickly build, fix, and figure things out with all the creativity in the group. It takes a team to build a robot.

Trouble Shooting

In an ideal world, robots would never break down and would function faultlessly. Troubleshooting problems cost time, and in the real world "money." When you learn the basic steps of troubleshooting and how to implement a critical thinking strategy, you reduce downtime (as in a robot competition) and frustration. You want your mechanical robot and software program to sync perfectly together to get your robot mobile. Troubleshooting is a trial and error system of checks and balances. It is valuable to consult with team members, rebuild, and sometimes reprogram the robot.

Virtual Worlds

Robot Virtual Worlds are exceptional online and (some offline) downloadable programs available to teach you RobotC and robotics. The unofficial slogan for Robot Virtual Worlds (www.robotvirtualworlds.com) is "No Robot, No Problem." Models can be built in software programs such as ModKit, AutoCad, and LEGO® Digital Designer. Practicing with 3D robotics without a physical robot is a great way to get hands-on lessons and practice.

Applying Math Concepts

In addition to building the robot, programming and learning to use remote controls, you will need some mastery of mathematics to get your robot moving. Math concepts can be applied to your robot depending on what you would like it to do. Simple machines will need to be programmed to perform a particular task. The relationships between different laws of force and motion and their impact on objects—including key factors such as momentum, friction, speed, acceleration, inertia, mass, balanced and unbalanced forces—will get your robot navigating with ease. These math tools work great for the Robotic Soccer Competition RoboCup. You can watch a game online.[5]

Understanding how changing gear ratios affect force (torque) and motion (speed) and their interrelationship will get you some good scores in a robotic competition. Calculate the robot speed using revolutions-per-minute (RPM), tire diameter, circumference, and consider the principals of friction, you will get your car moving at its fastest speed.

Robot Algebra, a project of Carnegie Mellon University and University of Pittsburgh's learning and Research Center ("Home", 2016) is a great resource for helping students get robots up and running in a fun and engaging way while applying algebra to the lessons. The robot can even dance with the right build and applied mathematical principles. A dance program can synchronize the linear movements and turns of each robot in a pattern. You use as several robots, and this example of how to program each robot to dance.

---

[5] https://www.youtube.com/watch?v=otsQL4SFxX0 retrieved July 22, 2016

Robot Dance		Robot 1		Robot 2 Motor values same as Robot 1		Robot 3 Adjustments synchronize moves	
Move Distance	Target	Motor Rotations	Measured Distance	Motor Rotations	Measured Distance	Motor Rotation	Measured Distance
Forward	50cm	5.3	50cm	10.61	9.35 cm	5.68	50 cm
Point turn right	45 deg.	0.71	45 deg.	0.71	129.8 deg.	0.25	45 deg.
Forward	50cm	5.31	50cm	5.31	93.3 cm	2.84	50 cm

There are just three movements in this program, and each is repeatable to form a dance sequence. The number of motor rotations and the measured distance determine the robot dance calculations. The third column shows the results the student will obtain if they program Robot 2 using the same values as Robot 1. This table shows distance travelled. You can use wheel diameters (different sizes) to change the speed the robot will travel. Speed is a proportional relationship. Other factors such as turning angle and rate form proportional relationships to complete the Dance Sequence.[6]

Summary

Robots are integrated into entertainment, industry and manufacturing, exploration, medicine, and in homes. Robots are designed and programmed for many reasons and are always ready for new transformations in the modern world. Becoming familiar with robotics in the early educational years will help you understand robotics in field technology, and possibly gain an interest in a career in robotics. In the next chapters, you will be able to get an inside view of the electronics and the mechanics that make robot controllers work, while learning about the math and programming that makes a robot intelligent.

---

[6] Home. (2016). Robotics Academy. Retrieved 29 December 2016, from
http://www.education.rec.ri.cmu.edu/content/educators/research/robot_algebra/)

References

Home. (2016). Robotics Academy. Retrieved 29 December 2016, from
http://www.education.rec.ri.cmu.edu/content/educators/research/robot_algebra/)

Picture of the VEX IQ brain. (2016). Pixabay.com. Retrieved 29 December 2016, from
https://pixabay.com/static/uploads/photo/2015/09/21/19/21/atx2-950511_960_720.png

Radio. (2016). En.wikipedia.org. Retrieved 29 December 2016, from https://en.wikipedia.org/wiki/Radio

Scratch - Imagine, Program, Share. (2016). Scratch.mit.edu. Retrieved 29 December 2016, from
https://scratch.mit.edu

The Intelligent Robot. (2016). I.ytimg.com. Retrieved 29 December 2016, from
https://i.ytimg.com/vi/tozwO1ZAzAg/maxresdefault.jpg

Robot Rules - Science AZ. (2016). Retrieved 29 December 2016, from https://www.sciencea-
z.com/content/widget/StuRobot_NewsArticle_6_L.pdf

CHAPTER 2

Types of Robots and Basic Components

In this chapter we investigate the mechanics of robots, components, and their applications. I am sure that you may have purchased many LEGO® Kits in stores; however, LEGO® has a division of technic bricks that are mainly used in high school robotics. I have worked with many students whose parents say that they have a room full of LEGO®, yet only a few students say that have used technic bricks, or seen a LEGO motor and battery pack. These technic bricks are easy to build with and are adaptable with gears axles to make mobile robots. We use these in the elementary school robotic programs as basic skill building enrichment classes to help students gain experience and confident in middle and high school robotics.

The types and components of robots will be discussed first because, that is the foundation of building, programming and controlling robots. Engineers and Programmers have made learning about robots easy, due to the simplified drag and drop graphical software programs that are available today, however; first came the basic components available to build robots. This will be explained in detail, before we look at the kits available for schools and homeschool use.

The three top robot kits used in middle/high schools are LEGO® NXT, EV3®, and VEX IQ®. There are other great models and kits available, however, we will discuss these, because they are the top kits used in National and International Competitions, and have many available resources including Virtual World Simulations. You can visit online and learn to build and program robots, without actually purchasing a kit.

Can You Name Some Robots?

Are you familiar with robots around the house? Surveillance cameras, automatic pool cleaners, sweepers, the infamous Roomba, vacuum cleaners, automatic lawnmowers, and any other robot that can complete a chore are several examples of robots used in the home.

Industrial robots carry a lot of clout. They usually have articulated arms that can accurately handle human functions such as welding, painting, and material handling, with precise programming parameters so that they can carry, lift and transport materials. They have been "programmed" to work autonomously in an environment specified by their design capabilities. Remember B & R Automation's golf ball assembly line? Arms pick and place items in the appropriate spots on an assembly line. Have you ever seen a robot that roams the factory floors carrying parts and orders to the workers?

Would you consider a computed tomography (CT) scanner a robotic assistant? Magnetic Resonance Imaging (MRI) is a scanning medical robot machine built for the purpose of performing and assisting in image-guided interventions. It is a sophisticated machine complete with rotating mechanisms that rotate 360 degrees around the patient. At numerous points, the x-ray beam records x-rays absorbed by the patient's body. The computer program is written by programming professionals to compute pixels of the body while the movement synchronizes to work with the remote sensing of the apparatus.

The Military has several types of robots; some dispose of bombs, transport, fly in areas with cameras (drones), explore the deep sea, and often search and rescue. Robots used for space exploration are amazingly designed. Just think of the types of materials that they need to be made from to make it through space. Phoenix Mars Lander has numerous robotic features, with arms and lifts to complete all the tasks to give scientists new information about Mars. Phoenix's job is to retrieve soil and ice as it digs through Martian grounds, and sends the samples to onboard scientific software and instruments that will analyze the samples. A good video can be found at https://www.youtube/phoenixmarslander. Last, we have entertainment robots, hobbyist and competition robots. Most of these will come with remote controls.

Motors and Mechanisms

A website to visit that displays many servo motors, DC motors and actuators is www.RoboShop.com. They have a nice variety of these motors and mechanisms. We will look at a few to discover the internal makings of a robot. Under the community tab on this website, is a resource for making **Arduino Robots**. Arduino is a type of robotic platform based hardware and software. It is open source (software for which the source code is made freely available and may be redistributed and modified), and the boards are similar to the intelligent brains of robots, just like the programmable brick brains that are prebuilt. You tell it what to do by sending a set of instructions to the microcontroller on the board. The programming language is software that is available for use with Arduino. There are tutorials and free interactive tools that will help you design your robot creation. There is also a site called "Let's Make Robots." If you visit the site, take a look at all the robots made by hobbyists.

Types of Actuators

Actuators are types of mechanisms for activating the control process of equipment. They are used in pneumatic, hydraulic, and electrical signals. Built inside the actuators are two separate modules, the signal amplifier, and the transducer. Amplifiers convert lower powers of the control signal to higher powers that are fed into the transducer. Therefore, the transducer's job is to convert the energy of the amplified control signal into work. Energy is converted from one form to another. We can appreciate the

Laws of Physics and Conservation of Energy. The **First Law of Thermodynamics** states that *"you can't create or destroy energy, but you can convert it from one form into another."*[1] Pretty much everything that happens in the universe obeys this fundamental law. We can apply electrical motors to this law because the motors convert electrical energy into kinetic energy.

Degrees of Freedom

There may be several types of actuators in robotic arms (moveable mechanical joints). Actuators will need to have characteristics of accuracy, precision, and reliability to meet the medical and industry demands for robots with arms that give rise to the Degrees of Freedom (DOF) needed for a robot to function in a rotational pattern. Free body in space has six degrees of freedom. Three for position X, Y, Z (**transitional**) and three for orientation: roll, pitch, yaw (**rotational**).

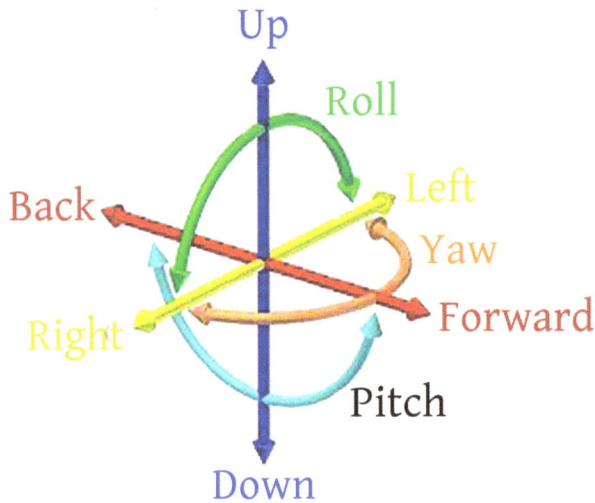

Permission to use by: GregorDS This file is licensed under the Creative Commons Attribution-Share Alike. 40. International license 2016

A robot arm manipulator has an end "efffector" with the DOF. Another example of DOF in 2007, Dean Kamen, (inventor of the Segway), unveiled a prototype robotic arm with 14 degrees of freedom. Life-like robots called **"Humanoid Robots"** can have 30 or more DOF, with five or six in each leg, six degrees of freedom per arm, and several more in torso and neck.

Types of Actuators

Robot Worx, *a Scott Company*[2] has listed many actuators and simple descriptions on their website. They describe the actuator as follows:

---

[1] Nesbit, B. (2007). Actuators. Handbook of Valves and Actuators, 279-310. doi:10.1016/b978-185617494-7/500419

[2] Types of Robot Actuators-Industrial Robots for Sale (06/28/2016) Retrieved from http://wwlrobots.com/education/actuators

Synchronous Actuator - The motor contains a rotor that rotates in synchrony with the oscillating field or current.

Asynchronous Actuator - This motor is designed to slip to generate torque.

Pneumatic - Powered by the conversion of compressed air, these actuators are used to control processes that require a quick and accurate response, but not a large amount of force. These compact and lightweight actuators are less energy efficient than electric motors.

Hydraulic - With the ability to convert hydraulic pressure and flow into torque and rotation, these actuators can be used when a large amount of force is needed. The most common example is a piston. This motor uses hydraulic fluid under pressure to drive machinery. The energy comes from the flow and pressure, not the kinetic energy of the flow.

Brushless DC Servo - This synchronous electric motor features permanent magnet poles on the rotor, which are attracted to the rotating poles of the opposite magnetic polarity in the stator creating torque. It is powered by a DC that has an electronically controlled commutation system instead of a system based on brushes. Current, torque, voltage and rpm are linearly related. The advantages of a brushless motor include higher efficiency and reliability, reduced noise, longer lifetime (no brush erosion), elimination of ionizing sparks from the commutator, and an overall reduction of electromagnetic interference (EMI).

Stepper - A type of brushless servo motor, this motor is electric and moves or rotates in small discrete steps. Stepper motors offer many advantages, such as dual compatibility with both analog and digital feedback signals. They can be used to easily accelerate a load because the maximum dynamic torque occurs at low pulse rates. Drawbacks of their use include low efficiency; much of the input energy is dissipated as heat, and the inputs must be matched to the motor and load. The load should be carefully analyzed for optimal performance. Damping may be required when load inertia is exceptionally high to prevent oscillation.

Brushed DC Servo - The classic DC motor generates an oscillating current in a rotor with a split ring commutator, and either a wound or permanent magnet stator. A coil is wound around the rotor, which is then powered by a battery. The rotational speed is proportional to the voltage applied to it, and the torque is proportional to the current. Speed control can be achieved by applying tape to the battery, varying the supply voltage, resistors, or electronic controls. The advantage to using a brushed motor over a brushless is cost. The brushless motor requires more complex electronic speed controls; however, a brushed DC motor can be regulated by a simple variable resistor, such as a potentiometer or rheostat. This is not efficient but proves satisfactory for cost-sensitive applications.

Traction Motor - A type of electric motor used to power the driving wheels of a vehicle. The availability of high-powered semiconductors has now made practical the use of much simpler, higher-reliability AC induction motors known as asynchronous traction motors.

AC Servo Motors - Used in applications that require a rapid and accurate response, these motors are two-phase, reversible induction motors that are modified for servo operation. AC Servo motors have a small diameter and high resistance rotors. This design provides low inertia for fast starts, stops, and reversals. AC Servo Motors can also be classified as asynchronous or synchronous ("Types of Robotic Actuators", 2016).

Alternating Currents and Actuators

Unless you are taking electronics in school, you may not use these types of mechanisms. As mentioned, most school robots like EV3 and VEX IQ come already made, but there is an advantage to knowing some of the mechanisms built into the robots and the main components. It might help you appreciate the work in electronics that paved the way for modern mechanical advances in technology with robots. The last chapter contains the mathematic principles that will be applied to your robot to help you make your robot move.

AC motors operate with a three phase current. You most likely will not use AC motors, unless you are doing something stationary like a crane lift, or robot arm. A DC (**direct current**) motor will have single flow. A direct current is the "Muscle Power" the converting DC and electrical power into energy force. Most common types of motors rely on the forces produced by magnetic fields.

Example of Motor Types and Their Applications

Motor	Speed	Torque	Application
DC	very high	very high	gates, trams, trains, bridges, hoists
DC Shunt	low	medium	the printing press, machines, conveyors
DC Compound	medium	high	shears, crushers, punch press

Motor	Speed	Power	Application
1 Phase AC	constant in operational range, slightly less than synchronous speed	3KW or less	medium power pumps, fans, and blower machines
3 Phase AC	nearly constant over large range of loads	High> 1.5 KW	high power pumps and machinery

The current motor needs have a **Stall Current.** It is held (stalls) so that it does not rotate. **Operating Current** will draw when a motor experiences "*zero resistance torque.*" Therefore, stall and operating current (maximum and minimums) are important to consider for the power you want your robot to have.

Voltage is *polarized*. That means current moves in one direction and cannot be reversed. The higher the voltage may mean more torque and will require more power. Mobile robots require more power, so it is rarely used with mobile robots. To optimize your robot, it will be best to establish curves relating to current, required torque, and voltage.

Velocity and Torque

The velocity of the motors allows for high speeds. Gears on a motor allow for the engine to run fast (Gearing Up). **Torque** determines acceleration. Therefore, a fast robot with poor acceleration is a slow robot. In class, we have had some fun challenges to see whose robot can go the slowest. We call it the Turtle Challenge. Gearing down and reducing torque will get your robot to win the challenge.

Caution: running motors close to stall current too often or reversing current can often under high torque, lead motors to melt. Motors are purchased with the stall and operating torque in mind, figure the minimum and maximum torque that you need for your robot. If you are not sure, consider torque before velocity.

Schematics Regulate the Robot's Power

Circuits control your robot's power source. It is a vital schematic to understand. **Power = Voltage x Current**. Batteries have different voltages for a purpose. It will not work if the power regulation circuit is not calculated correctly. Set up the circuit correctly, to get an efficiently running robot. There are some steps to follow, regulate at a set voltage, supply a minimum required amount of power at all times, and allow for additional special features and requirements to run your applications.

Optimally, it will be best to use a power source closest (and slightly above) the desired voltage input requirement. However, this is rarely easy or feasible. The reason may be because of the different electronics that require different voltages. A microcontroller will require 5V, your motors perhaps 12V, a voltage amplifier perhaps both 20V and -20V.

Battery Life Chart

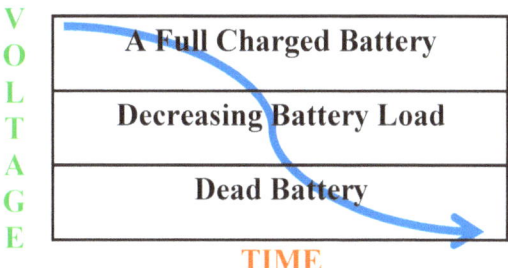

Sometimes, you may want to use rechargeable batteries for your robot. Batteries can be an expensive addition to running and operating a robot. The kits used for schools may come with rechargeable batteries. Because batteries run down due to voltage vs. time, batteries are never at a constant voltage they will drain. A 6V battery will be close to a 7V when fully charged, and can drop to 3-4V when drained. The image shows how a typical battery voltage changes over time.

A real good source for advanced robotics information about Schematics can be found at ("How to Build a Robot Tutorials - Society of Robots", 2016)[3]. Visit this site if you want to learn more about microcontrollers and sensors and voltage regulators. The site is maintained by John S Palmisano, a scholar in the field of Robotics. Here, advanced students can get extra help with schematics and mechanics of robots in the forum available to the robotic community.

Servo Motor with Batteries 6V Arundo Servo Sweep

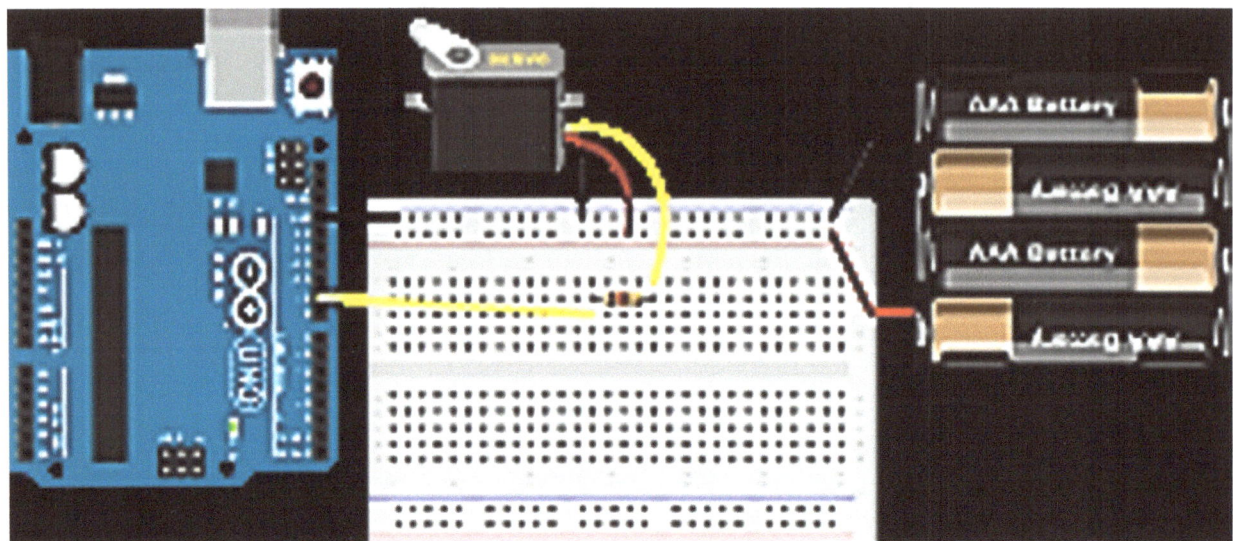

By Nevit Dilmen (Own work Made with Fritzing) [CC BY-SA 3.0 (http://creativecommons.org/licenses/by-sa/3.0)], via Wikimedia Commons

Sensors (Inputs)

Have you ever wondered about the intelligence of a Robot? Robots can react to their environments given the right information. Programs are written to give robots the sense of sight, sound, touch, smell, and taste. The same types of information we receive through our brains about our environment. Sometimes they can be programmed to have a better insight (sensing) into information that is beyond our human capabilities. Like a bat or barn owl, we can program a robot to "see in the dark," and detect minute amounts of invisible radiation, or measure movement too small for the human eye to see.

When you learn about programming in the next chapter, you will be able to tell the robot what to do using different sensors that attach to the robot. Physically, the robot will respond to distance, light, sound, temperature, pressure and in more advanced robots, altitude, and contact. What is some technology associated with sensors in robotics? Pressure gages, gyroscope, compass, microphones, cameras, infrared, and radar, are all types of sensors used in industry.

---

[3] http://societyofrobots.com

Picture of Sensors

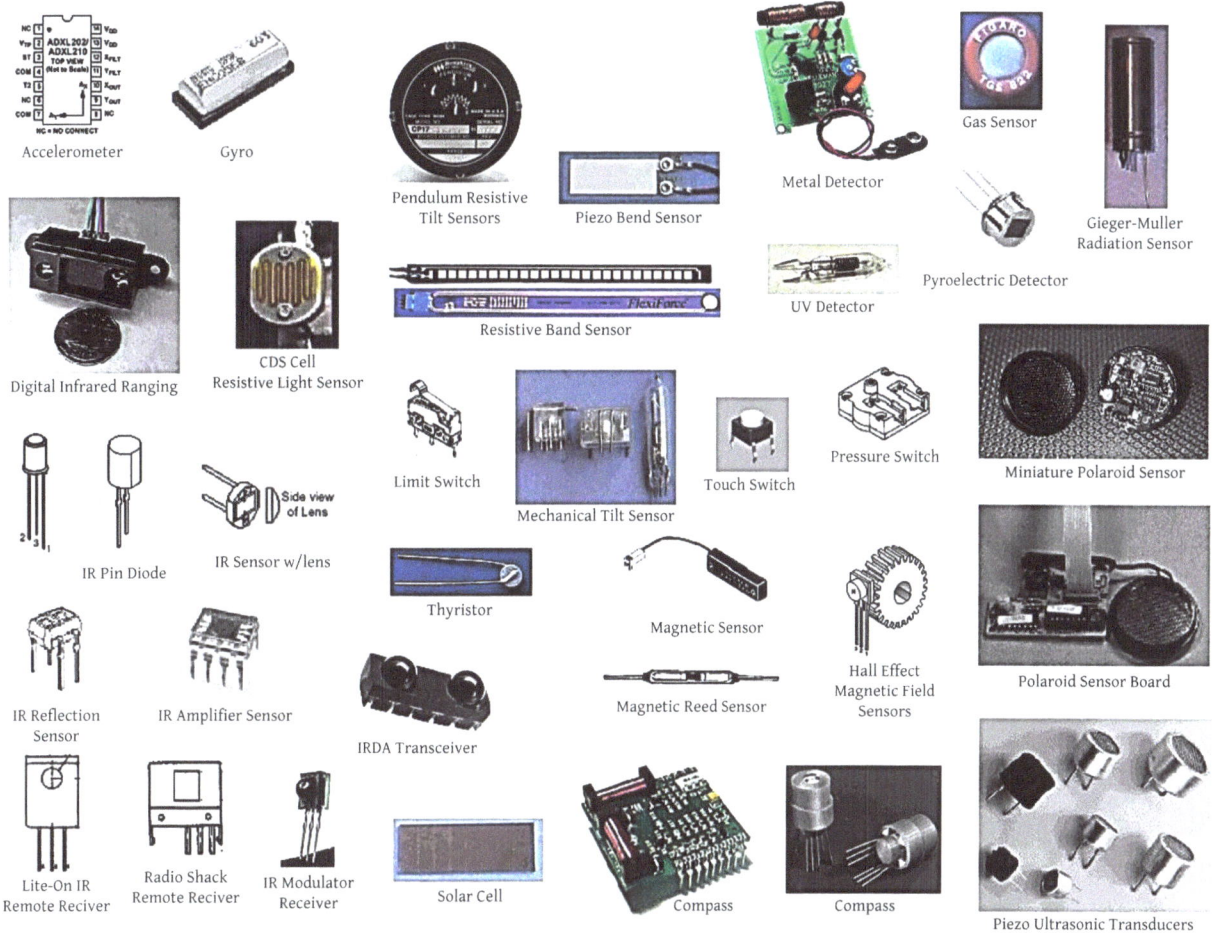

("Types of Sensors by Electrical4u", 2016).

Light Sensors

Light sensors measure the amount of light that can impact a photocell. It is a **resistive** sensor. When brightly illuminated, the resistance of a photocell is small. The resistance of a photocell is low when it is brightly illuminated and *visa-verse*. In that sense, a light sensor is reading the "dark" sensor.

Infrared (IR) Sensors

Another type of light sensor is Infrared sensors which function in the infrared part of the frequency spectrum. IR sensors involve an emitter and a receiver and are considered active sensors. They can be understood the same way that visible light sensors work, as motion detectors (break-beams) and as reflective sensors. In robotics, we prefer to use IR and other applications because it suffers a bit less from ambient interference (sunlight, incandescent light, electromagnetic field, and fluorescent lighting) and it is easily modulated, and simply because it is not visible.

Infrared (IR) Communication

IR modems work with modulated infrared in a **serial line** for transmitting messages. The two methods are **bit frames** and **bit intervals**. Frames sample the middle of each bit; assumes that all bits take the same amount of time to transmit, and intervals (common in commercial use) sample at the falling edge, duration of intervals. Sampling determines whether it is a Zero or One.

Ultrasonic Distance Sensing

One sensor fun to program and play with is the NXT Ultrasonic Sensor. It can gives life to your robot. Ultrasound sensing is based on the **time-of-flight** principle. How it works is that the emitter produces a sound. It is a sonar "chirp" of sound, which travels away from the source. It can encounter obstacles, reflect from them and return to the receiver (microphone). You can track the amount of time it takes for the sound beam to come back by starting a timer when the "chirp" sound is produced, and stop it when the reflected sound returns. This is a simple example of how to compute the distance the sound travelled. **A useful constant to remember is that at room temperature, sound travels at 1.12 feet per millisecond, or 0.89 milliseconds per foot.**[4]

Echolocation is the process of finding one's location based on **sonar**. The ultrasound sensor was designed work with echolocation. Just like bats, ultrasound is used instead of vision at night in their caves. Bat sonars compared to artificial sonars, are much more sophisticated. Bats have higher frequencies and can find the smallest, fastest prey while avoiding all the other bats in their community.

Touch Sensors

There are a few varieties of touch sensors that come with the robot kits that you will use in the classroom or at home. Lego® NXT has a touch sensor that can be programmed to stop and turn if it is pressed. Students will build a type of grill that will be pressed, or it can be used at the end of a wire as the touch point, you can use your thumb to press it. Build the walking dog that can bark and turn its head with this touch sensor ("nxtprogram.com", 2016). It one of the many robots designed by Dave Parker, NXTPrograms.com. In the picture depicted, the ultra sonic sensor and the sound sensor does not work. It is just used to show a puppy like robot.

Permission to use by: Dave Parker, NXT Programs, 2016)

LEGO® EV3 has a touch sensor. It works similar to the LEGO® NXT. It is an analogy sensor that detects when the front button is pressed or released. The block for the code can be programmed to press or release upon touch. It can count single or multiple presses. One fun use of the touch sensor is to program the

---

[4] http://www.ministereo.com

robot to navigate a maze. If it hits a wall, it can turn and try to continue to find its way through the maze. Similar sensors are built into the technology of digital musical instruments, and computer keyboards.

Vex IQ has a touch sensor with some advanced features. It comes with a default program meaning it can be attached to the robot and used right out of the box. The input of the sensor adds some interesting dimensions for your robots interaction and programming. The LED can change. It can sense the actual colors it is reading "seeing" and can be programmed to change colors like changing the colors of eyes. It will work like an on-off switch using the remote controllers. Attach it to your robot and program it to change red and green to show that your robot is turned off. The Touch LED is a useful sensor that can be customized for your robot. It can stimulate robot eyes, or just to make them a special robot. Modes, (a status of a button or the state of a software function) can be programmed to show a certain color. These three colors will glow for the specific program.

Vex IQ Touch LED Changes Colors
Creative Commons Attributions 3.0 Un-ported License

Have you ever seen the "clapper" lamp commercial for the lamp that you can turn on or off by just a clap? Here are some explanations. You can program LEGO MINDSTORMS NXT robots to move at the sound of hand claps. Here are few concepts to understand about sound and how it relates to sensor functions. A review of the human ear is a good place to start, let's consider the ear and links it to a robotic sound sensor and its applications.

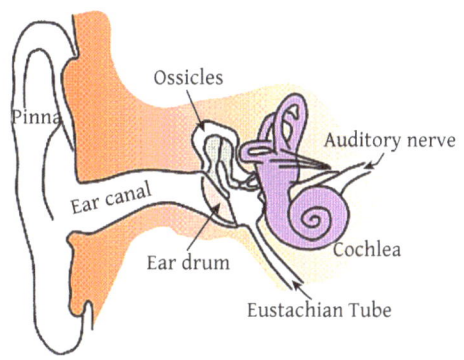

By Iain at English Wikipedia[5]

The ear pathway has an eardrum that will vibrate when it receives sound. It is air pressure vibrations that are converted to electrical signals and then transported to the brain. Sound sensors have been created by engineers to emulate this human feature. Like the LEGO® NXT sound sensor, it works like a human experiences sound. The sound sensor has a diaphragm (similar to an eardrum) that converts air pressure vibrations into electrical signals that are transported to the NXT brick/computer. Microphones work this way, too. Sound sensors are not usually part of competitions, because they do not work as well due to the noise in the competition arena.

A circuit board for sound   *The NXT Sound attachment*

Sound Sensor Readings	4-5%	5-12%	12-34%	34-100%
Similar Sounds	Silent Room	Person Talking Far Away	Person Talking Nearby	Shouting or Playing Music

You can calculate the threshold by using this formula **5 + 95 = 100 100/2 = 50** This is the Threshold Value. The value will be able to programed into your robot sound sensor block.

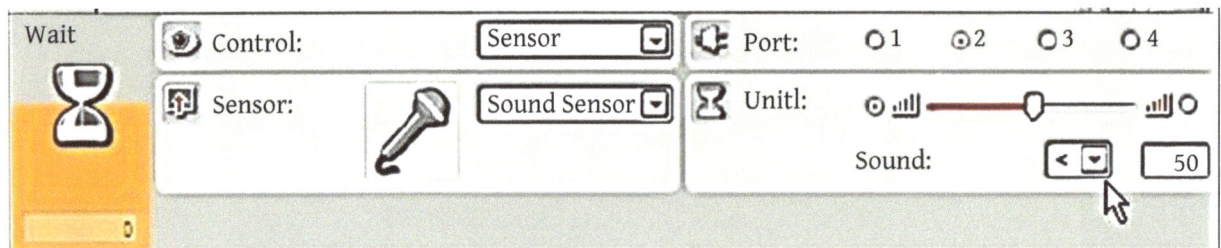

We have two ears which enable us to hear sounds. Example: **Sound waves from a siren>human ear>human brain>Heart> sympathetic nervous system>move your car over!** This is an example of the "stimulus-sensor-coordinator-effector- response flow chart.

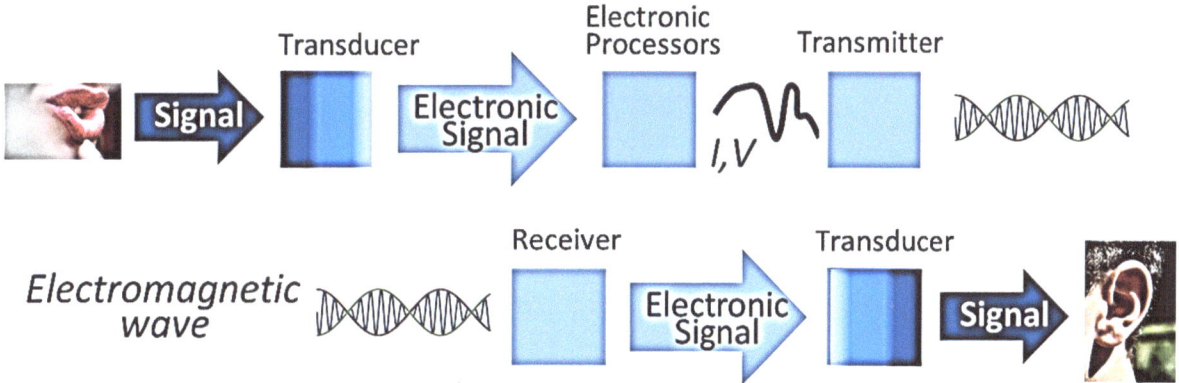

Other examples of engineering systems that use sensors and are similar to the human ear are the LEGO® sound sensor, computers with sound cards, karaoke machines, microphones and mics in cell phones.

We can sketch out the stimulus response sequence for a robot sound sensor using a LEGO® sound sensor. The sensor will "read" the program and send it back to the brick computer brain. It will act on the information it receives from the sensor. The sensor will send the information through the wires (similar to our nervous system pathway) and connect them to the brick the information gathered will have the robot respond based on what you want it to do.

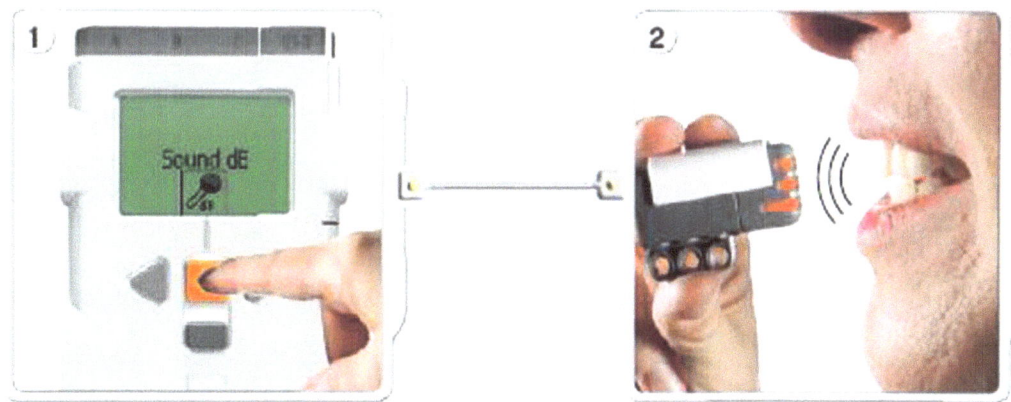

Picture adapted from LEGO® 9797 Users Guide 2006

Example: Below are a few code blocks from NXT software program. The sound sensor operating modes measure, compare, and calibrate. Decibels(dBA) measure the sound level pressure. The threshold of values found in the NXT sound icon are 0-100 percent that tunes and filters the noise level to mimic the sensitivity of the human ear. Comparing and measuring differences with a numerical value, allows you to select the value in which your robot will respond to the noise of the clap.

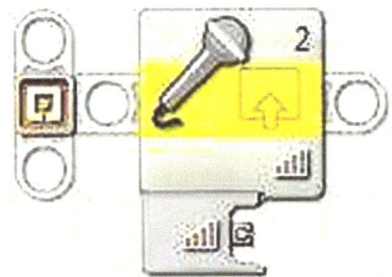

Features- can not be used in competitions due to too much noise at the competitions. **Yellow block**-reads the volume and **Orange block**- waits until the sound is at a particular dBa given an event. Threshold volume can be greater or lesser. It is Attached at least 15mm from the motors so that it does not pick up the noise of the brick brain.

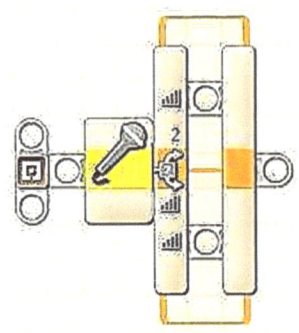

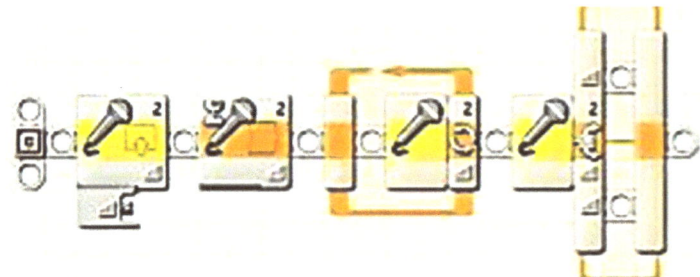

Switch Block

This program will make the robot start on clap and then stop on clap. There is a common problem with the sound sensor; the microcontroller may hear the program as two sounds. The solution- be sure to silence for 1 second for the program to work. Program in one second in the Next block and the robot waits for the clap. Then starts and stops upon clapping.

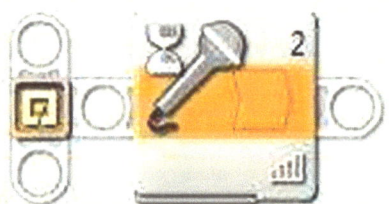

Temperature Sensor

LEGO® has a digital temperature sensor and works with EV3 platforms as well. It is a digital temperature sensor and measures both Celsius and Fahrenheit. The older thermometer was analog and was not near as precise as the newer digital temperature sensor. Why might you want a temperature sensor to attach to your robot? You may want to track some data for a project such as how temperature changes with the depth of water, or at very specific points on a surface.

In this blog, Damien Kee, Robotics Educator and author of several books, "*Guides for Busy Teachers*" has used a similar sensor "Dexter," and it will work with the NXT brick brain. It is an excellent example and a few pictures that explain how the temperature sensor works can be found on his website ("Damien Kee - Home", 2016). (SEE REFERENCE)

The Role of Gears in Mechanical Devices

Gears are important for our robots because pairing the gears together makes a robot mobile. Similar to bikes that come with simple gear trains, to more advanced for mountain biking. The gears are made to work harder, so the all the work is not in your legs. A simple bike has gears and a pulley. This is a gear-train. Engineers need to understand the relationship between the gears, the outputs, about the change of torque and speed, and direction of the power source.

Gears are wheels with teeth that mesh together with other gears and are usually made of metal or plastic. Examples: car transmissions wind-up toys, non-digital clocks, drills, bicycles, powered wheel, chairs, and lifts.

Mathematically gear ratios are calculated like this:

Gear ratio = $\dfrac{\text{Effort Output Gear Number of Teeth}}{\text{Load Input Gear Number of Teeth}}$

Imagine two 24-tooth gears contacting each other. When one 24-tooth gear turns once, how many times will the other 24-tooth gear turn?

Now, imagine replacing one of the 8-tooth gears with a 40-tooth gear, with the 40-tooth gear turning the 8-tooth gear. *Calculate when the 40-tooth gear turns once, how many times will the 8-tooth gear turn?*

Your answers should be one time and five times. The large gear turns slowly it has more torque (rotational force). Sometimes when we are programming our robot brick brains, we need to remember that the power of the machine may need to be reduced or speeded up. Sometimes your robot needs to be reliable or fast. Using 100% power in the parameters will not be as efficient unless you apply the design for one or the other. It is important to calculate speed or torque (strength). **Power = torque x speed**. If the EV3 or the Vex IQ motor power setting is fixed, you will need to remember that increasing the torque decreases speed, and if you increase speed you will decrease the torque. The larger the rotational force, the larger the gear ratio. Smaller gear ratios produce faster speed.

A Gear Ratio <1(less than 1) is called **Gearing Down**. There is always some loss of friction. Will the speed of the **wheel** go up or down compared to the **motor**? Gearing <u>Down</u> = <u>Speed</u> goes <u>Down</u>, but, power (torque) increases. Why might you want to do this?

A Gear Ratio >1 is called Gearing Up. Will the speed of the **wheel** go up or down compared to the **motor**? **Gearing <u>Up</u>** = <u>Speed</u> goes <u>Up</u>. But, power (torque) goes down. There is always some loss to friction.[6] Why might you want to do this? Want the short answer? Gearing down by installing a larger gear increases the final drive ratio and reduces top speed, but can increase acceleration. Gearing up, like with a smaller rear gear, decreases the final drive ratio and adds more top speed to a motorcycle or ATV.

---

[6] Two Gear Gear Trains http://slideplayer.com/slide/8675853

If you have a robot built at home or school, try to visit ("Dr. Graeme - Free Lego MindStorms NXT tutorials", 2016) to explore a gear challenge "Climb the Highest Mountain." Lots of gears and explanations for you to explore with Dr. Graeme ("Free Lego NXT Mindstorms Robotics tutorial Climb A Mountain Use of gears Challenge 30", 2016).

Other notable gears transmit rotations in perpendicular directions; their axles are perpendicular, similar to **bevel gears**. Often robots are built with locking mechanisms, so that the rotation is only transmitted in *one direction*, from the worm gear to the spur gear. Belt and pulleys are part of the simple machines that you studied in elementary school; they are ratios of diameters. The transmit motion across a long distance and rotate in the same direction (this is opposite of two gears). Two gears together work clockwise and counter-clock wise as the teeth mesh together.

Guitars and other stringed instruments have worm gears for tuning. Worm gears are found in the machinery of elevators. Large trucks and off-road vehicles like a Hummer often need to deliver different amounts of torque to each wheel, depending on what action the vehicle wants to perform. One good thing to remember about gears is ***never to drive a smaller gear with a larger gear***. It can risk exceeding the DC motor torque and rating, and damage the inner gearbox.

Arms

In the first chapter, we discussed why we need robots. They do not get tired, work with precision and their only needs are mechanical in nature. That can be fixed. Industrial robots that use mechanical arms designed for use in the industrial robots do jobs that are difficult, dangerous or dull. These robot arms take the place of human jobs by lifting heavy objects, painting, handling chemicals, and performing tedious assembly work.

The main categories of industrial robots with some description of their mechanical structure are:

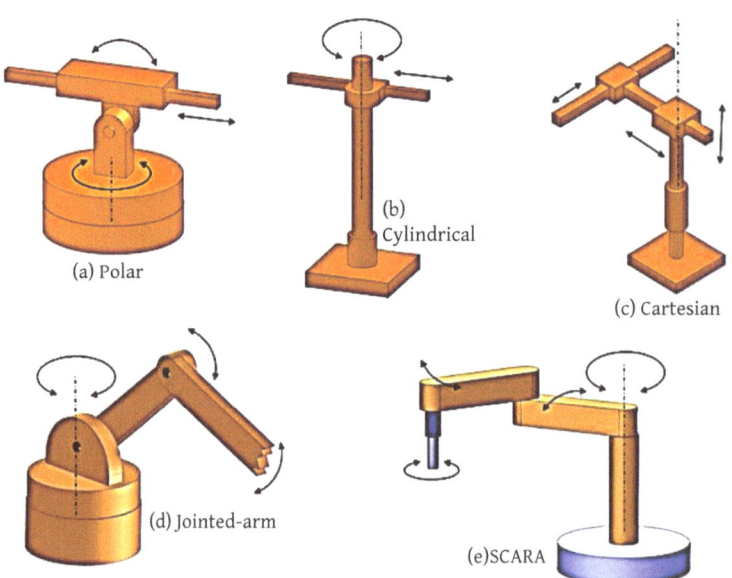

(a) Polar

(b) Cylindrical

(c) Cartesian

(d) Jointed-arm

(e) SCARA

- Cartesian robot/Gantry robot: Used for picking up and placing, applications are assembly operations, handling machine tools and arc is equal to a Cartesian coordinator.

- Cylindrical robot: Used for assembly operations, handling at machine tools, spot welding, and handling at die-casting machines. It's a robot whose axes form a cylindrical coordinate system.

- Spherical/Polar robot: Used for handling at machine tools, spot welding, die-casting, fettling machines, gas welding and arc welding. It's a robot whose axes form a polar coordinate system.

- SCARA robot: Used for pick and place work, application of sealant, assembly operations and handling machine tools. It's a robot which has two parallel rotary joints to provide compliance in a plane.

- Articulated robot: Used for assembly operations, die-casting, fettling machines, gas welding, arc welding and spray painting. It's a robot whose arm has at least three rotary joints.

- Parallel robot: One use is a mobile platform handling cockpit flight simulators.[7]

Wheels-Let's Move that Robot!

One good thing about my LEG® kits is that I have several sizes of wheels, it came with collecting LEGO parts, and the tires were thrown into the bags. Students like to grab the biggest tires, and I use that as a learning curve. "We will see," I say or "think." I allow them to try them out to see if the overall design of the robot will work to with the program and the wheels. Many times they want to change them out for the smaller wheels. Robots are pretty cool once you get the right size wheels.

When we program in the next chapter, you will learn that the basic robot is a differential drive. The movement of rolling around on the ground is controlled by sending different amounts of power to the left and right wheels. Both wheels are powered by different amounts, and one wheel covers more ground that the other. An example of this is that if the left wheel is rotating faster than the right one, the robot wheel will turn left while traveling forward. To travel straight forward, you may have to apply the same amount of power to both wheels in the forward or backward direction.

Try this with your robot. Attach a pen to the robot, and the robot wheel will rotate around its center. It can hold a marker and draw a circle. Practice with the pen attachment, changing out the degrees, rotations and change steering to see your robot draw on paper. The graphical program software that you will be working with will allow you to program in rotations or degrees.

---

[7] Mark Tyson, ABB Technology Show, May 2013 Robotics 101. (May 07, 2016). Retrieved from http://www02.abb.com/global/zaabb/zaabb011.nsf/bf177942f19f4a98c1257148003b7a0a

Driving wheels are weight-bearing wheels not driven by motors. They are caster wheels; a special kind of driving wheel that allow wheels to swivel left and right eliminating friction when the robot turns. It is easy to build and attach to the back of your two-sided robot. Now you can build your robot!

Castor Wheel using LEGO® Technic Parts

References

Types of Robotic Actuators. (2016). Robots.com. Retrieved 30 December 2016, from https://www.robots.com/education/actuators

How to Build Robot Tutorials - Society of Robots. (2016). Societyofrobots.com. Retrieved 30 December 2016, from http://www.societyofrobots.com/schematics_powerregulation.shtml

Types of Sensors by Electrical4u. (2016). Electrical4u.com. Retrieved 30 December 2016, from http://www.electrical4u.com/images/sensor-2-16-01-14.jpg

nxtprogram.com. (2016). Nxtprogram.com. Retrieved 30 December 2016, from http://www.nxtprogram.com

Damien Kee - Home. (2016). Damienkee.com. Retrieved 30 December 2016, from http://www.damienkee.com/

Free Lego NXT Mindstorms Robotics tutorial Climb A Mountain Use of gears Challenge 30. (2016). Drgraeme.net. Retrieved 30 December 2016, from http://www.drgraeme.net/DrGraeme-free-NXT-G-tutorials/Ch30/Ch30V1BCG/default.htm

Dr. Graeme - Free Lego MindStorms NXT tutorials. (2016). Retrieved 30 December 2016, from http://www.Drgaeme.net

CHAPTER 3

Types of Controllers and Applications

Once you understand types and components of robots, you will be able to gain more insight into many tasks they can perform, and some functions entrusted to them for the particular tasks they perform. Controllers fall into two main categories, the **telerobot**, and the **autonomous robot**. The telerobot operates via remote control operated by a human, while the autonomous robot will be operated using **logic control** and **behavior control**. The telerobot (semi-autonomous) will also work **tethered** (physically connected to the robot), or **wireless** (radio wave or infrared controlled). The autonomous robot will access a set of codes you have programmed into the microcontroller's brick brain via software programs and carry out the coded instructions without any human intervention, or interruptions.

How to Choose your Telerobot

For the record, there are many types of controllers, joysticks, robot kits and software. I cannot explain them all in this chapter. Listed below are some of most popular for after school programs, competitions, and educational robots teachers may use for the classroom. The NXT Mindstorms®, the VEX IQ® and the EV3® all come with **"drag and drop"** graphical software, and they also work with ROBOTC®. NXT Mindstorms being the oldest and many kits are still in use. However, NXT does not do any updates to the firmware, since the new launch of the EV3 robot. The NXT price is not as high-priced, since the other model kits came available, and it does much of the same things and the EV3. You can find used kits and "Brick Brains" on EBay.com, just be careful to note the amount of use they may have.

There is an available website www.nxtprograms.com, managed by Dave Parker, and it is like a treasure chest full of models, rbt. files, and full explanations on how to use and build NXT and now EV3. The LEGO® remote control paired with the NXT Infrared receiver (IR) HiTech (NIR1032) will work with the NXT Brick, but you will only be able to navigate your model forward and backward. It is still a fun way to attach them and use with your NXT robot. VEX IQ has forums for discussions that can build upon skills and knowledge in real time, because you may find a seasoned robotic instructor who will answer your questions.

OVERVIEW			
microcontroller (Brain)	VEX IQ	EV3	NXT
Year Released	2013	2013	2006
How are they used?	Building functional robots competitions	After school, Technology class, and works with all LEGO Technic Bricks.	
Purchase for personal use.	Vex is right for schools and home building. It has sensors that are more sophisticated than the other brands. There game challenges for competitions such as VEX High Rise.	If you already own other LEGO parts (mainly TECHNIC), you can use them with the MINDSTORMS sets	
HARDWARE			
IO Ports	12 either motor or sensor	4 motor/4 sensor	3 motor/4 sensor
USB Connector	micro USB	mini USB	Standard USB
MOTORS			
Can be used as servo	Y	Y	Y
Torque control	Y	N	N
Small and easy to mount	Y	N	N
DISPLAY			
Resolution	126x64	-	100x60
Backlight	Y	N	N
Buttons	4	6	4
SENSORS			
Color/Light	Y	Y	Y
Ultrasonic (distance)	Y	Y	Y
Gyro	Y	Y	N
Touch	Y	Y	Y
Touch-LED	Y	N	N
Other 3rd Party	N	Y	Y
IR Beacon Sensor (important for Robocop Jr)	N	Y	3rd party sensor
Batteries	Includes rechargeable battery	6x AA; rechargeable available	6x AA; rechargeable brick battery pack

			available with certain kits, or sold separately
WIRELESS			
Remote	400/900 MHz game controller	IR requires sensor port	NXT IR Receiver Sensor HiTech (NIR1032)
Bluetooth	Y You will need two Smart Radios (one for Robot Brain, one for Controller) to enable telerobot control to work with the wireless programming/debug features. You can use one Smart Radio in only the Robot Brain if you wish to have an autonomous-only robot. The Robot Brain and Controller must always use the same type of radio for them to communicate with each other; you can't mix one Smart Radio with one 900 MHz Radio.	Y	Y
Wi-Fi	N	With a USB dongle	N
SOFTWARE			
Graphic Language	Y	Y EV3 Free to download http://www.lego.com/en-us/mindstorms/learn-to-program	Y NXT
RobotC	Y	Y	Y
Included RC program	Y	Limited	N
Download Wireless	N	Y	Y
Brick-to-Brick Communication	N	Y	Y
On-Brick Programming (program without PC)	N	Limited Can be uploaded to PC later	Limited

Remote Controllers

Remote controllers can be built for a particular brand, yet there are ways to connect other equipment such as iPhone/iPad, Android phones, Xbox game controllers, PS4 game controllers, and even use two Intelligent Brick Brains as a controller and receiver. Let's start with an easy one to build and get operating quickly, the VEX IQ model with controller, because it has a default program that already comes with a ready, built-in controller and receiver program.

On the Vex Robotics website, you can find information about the VEX IQ® controller. It is a cool piece of equipment to use with your model build, and especially fun in competitions. The initial tutorial for VEX IQ has a model build plan and instructions to get your robot built, and running immediately. It will give you instant gratification and confidence to learn more about robots. It works similar to a joystick for video games. The custom controller is ergonomically correct for your hands. There are two joysticks with fine control that give some precise driving, turning, and raising mechanical arms. Extra buttons allow for additional controls such as opening and closing a claw.

How does the VEX IQ® Work?

The center of the controller is a microcontroller built by Texas Instruments MSP430. It reads the inputs (the codes you have written) and transmits them through the tether port and also wirelessly. It has 3 Ports, a tether port to wire communication to the Robot Intelligent Brain, USB port for charging and a radio port. The Radio Frequency is compatible with the 900 MHz Radio and is used in USA and Central America. It takes lithium batteries.

The joysticks are used for navigation and include power and speed. The buttons can be programmed to control a behavior you want the robot to do. Some behaviors that work with a claw mechanism would be to open/gripper and close/gripper. The buttons will control right, left and straight movement. The 12 programmable buttons are broken into two groups of four, each having up, down, left and right functions ("Bluetooth Radio - VEX IQ Forum", 2016).

Trouble Shooting the VEX IQ Controller

VEX IQ robotic system has made it easy for you to understand troubleshooting. In their troubleshooting guide, you can find easy instructional information for help with calibrating. The VEX IQ controller may need to be calibrated. Once the robot is up and running, calibrates the controller. Calibrating will give the robot more absolute accuracy, than a non-calibrated robot. Sensors can be calibrated on some robots. The calibration function should run by default. VEX IQ instructional guide suggests the way to tell if you need to calibrate is if you notice the robot drifting while navigating even when the joysticks are in the center position. The joysticks can be moved in different directions, and they work by changing the speed at which the motor will spin.

- Use the Tether Cable to connect the Controller to a Robot Brain Connect[1]

- Turn on the Robot Brain and the Controller

- Press the (X) button on the Robot Brain

- Select "Calibrate Controller."

- Rotate thumb-sticks 360°

- Press (✓) to save

From the instructional guide, you can follow the step by step calibration instructions: it will guide you and help diagnose a controller problem.

Operating the Controller Using ROBOTC Software

ROBOTC Graphical software can sync with the VEX IQ Remote Controller. It comes with two controls, tele-operator and autonomous. The VEX IQ brain needs to link up with the wireless remote controller, and with ROBOTC, you will be able to customize the remote control program but, with just a few codes the robot will be able to navigate.

1. Configuration

2. Repeat(forever)

3. Arcade Command

The drag and drop program uses a **continual loop** block that is synced with the data from the remote controller and activates the VEX IQ built robot. A standard Ethernet cable is used to "tether" the controller with the brain. It will sync and identify the robot, working only with your robot. The joystick axis, along with the buttons fine tunes the navigation and mechanism. "**Controller Resolution**" is the smallest increment a motion remote controller can sense. The joystick buttons have standards of 1-pressed, 0- not pressed/released. The joystick axis returns standards of -100 to +100. These values represent motor power outputs. Think about turning the volume up and down on the radio.

```
1  repeat ( forever ) {
2      tankControl (ChD ▾ ,ChA ▾ ,  10  );
3  }
4
```

---

[1] Controller - VEX Robotics. (2016). Vexrobotics.com. Retrieved 28 December 2016, from http://www.vexrobotics.com/vex-iq-controller.html

The 3rd box of the tankControl command block allows you to identify a "**tolerance**" for the VEX IQ Controller joysticks. Tolerance deals with the robot's accuracy of just how close the robot can achieve mobile positioning. The robot repeats over and over again, with the loop block.

```
1  repeat ( forever ) {
2      armControl (armMotor ▼ , BtnLUp ▼ , BtnLDown ▼ , 75 );
3      tankControl (ChD ▼ , ChA ▼ , 10 );
4  }
5
```

This robot has an arm attachment. The arms can be controlled using ROBOTC." **Button Up"** and "**Button Down"** will need to be part of the **loop configuration.** It allows for the arm move up and down with the buttons of the controller. The last block with "75" is the **speed** of the arm motor control. An important programmable feature needed for industrial robots designed to operate in a factory.

Other Controllers

The Xbox 360 and the PS4 Controllers work with Robot C. You can go to Robot- Debugger Window and click on the Joystick Control- Basic: This is a window that can also provide live feedback of your controller movements.

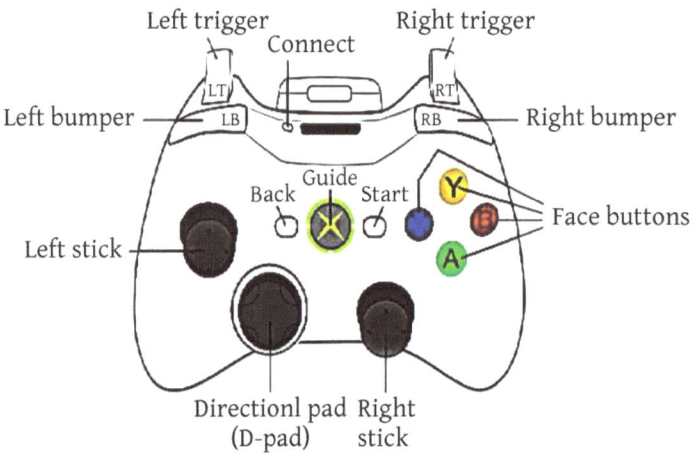

("*File:360 controller.svg - Wikimedia Commons*", 2016).

Infrared Receivers and Radio Wave

Infrared receivers are built to use with controllers and also with robotic sensors. The principle operation is infrared light which is reflected when hitting an obstacle. The **Infrared Receiver** captures the light reflection while the voltage (electrical energy) is calculated based on the quantity of light it receives. This can present a problem with some types of IR receivers used outside. They may not perform as well in sunlight.

Radio waves are all around us. They have long wavelengths and low frequencies. The radio wave is found on the electromagnetic spectrum. Remember red is the longest wave? Radio waves also have low photon energy. What does that have to do with robotic controllers? The radio transmitter uses the electromagnetic wave by means of low frequency to transmit data to the **Radio Frequency (RF)** receiver, which in turn is encoding information in wave patterns. Electromagnetic waves move at a "**speed of light**," becoming a good source to carry information from controller to receiver. The frequency of the radio wave is a "**carrier**" wave and carries the signal wave to the receiver up to approximately 200 meters.

The Radio Transmitter employs an electromagnetic wave by means of a certain frequency to broadcast information to the RF receiver module. **Frequency Modulation (FM)** is encoding data in wave models. Waves have numerous essential properties, including *amplitude* as well as *frequency*. In FM, the frequency of the fundamental wave is called a **transporter wave**, and is modified via collecting this by the other signal called the **modulating wave**. That creates waves which seem uneven although it is transporting the information from the signal wave on top of the transporter wave. Frequency modulation is less susceptible to interference than other radio transmission techniques, for example: direct transmission or **Amplitude modulation (AF)** of the signal wave. As electrons speed up, it emits electromagnetic signals which shift at a pace of light. Through speeding up and slowing down electrons in a controlled model, a wave model might be creating an electromagnetic field. This is known as electromagnetic waves. Electromagnetic waves move at a speed of light and that's why the waves are very good at transporting data rapidly from one point to another point.[2]

Nikola Tesla's Remote Controlled Boat

Back in 1898 in New York's Madison Square Garden, Tesla was promoting a new invention. A remote controlled boat. He put a small, radio-transmitting control box, together along with a propeller and rudder, attached an electric motor to drive the rudder as well as the propeller a storage battery. He designed a method for getting radio signals from the control box to move a propeller and rudders. He set up the exhibition with a small indoor pool and a 4-foot miniature ship with the control box by numerous levers. The deck had antennae attached for receiving signals and light bulbs *"Indeed few people at the time were aware that radio waves even existed and Tesla, an inventor often known to electrify the crowd with his creations, was pushing the boundaries yet again, with his remote-controlled vessel"* (Turi, 2014).

Tesla could control the speed and direction while displaying extra attraction features like turning on lights and other moving parts. He certainly amazed his audience with a wireless boat.

---

[2] Chapter 19 Electromagnetic Radiation, (2016)." Physical Science: Teacher Wraparound Edition. New York: Glencoe/MCGraw-Hill, 1999. 529-30 print.

("File:Tesla boat1.jpg - Wikimedia Commons", 2016).

Wireless Networks

Most wireless networks are designed with computer networks that use wireless data connections to connect to a network node. In homes and businesses telecommunication networks are set up with radio communication. We know these networks as a cell phone, satellite communication networks, and terrestrial microwave networks.

When working with robots or wireless controllers, we can use Bluetooth to get a longer range for our robots to navigate. Some recent applications that have been seen in an Article in the New York Times reports on July 8, 2016[3], in Dallas Texas, five police officers were slain and authorities decided to use a scout robot that was designed to go into building carrying detonating devices that would help to alleviate the need for human intervention in a hostile environment. The detonating device attached to the robot contributed to bringing the standoff with the assailant to an end.

A remote user interface built to control the robot via wireless technology must be able to control a wireless robot via Bluetooth without the need in direct sight as the Infrared Receiver does. The robot can be located behind some object or wall, and the communication will not be lost. You can still control the robot with a joystick or keyboard while the robot collects data as it navigates a path.

Remote Control Using Bluetooth

Some general types of electrical connection examples are a car lock and door opener, a garage with a garage door opener, and the connection between and iPod and a computer. The difference between a

---

[3] Fernandez, Manny, Richard PÉrez-peÑa, and Jonah Engel Bromwich. "Five Dallas Officers Were Killed as Payback, Police Chief Says." *The New York Times.* The New York Times, 08 July 2016. Web. 06 Jan. 2017. <http://www.nytimes.com/2016/07/09/us/dallas-police-shooting.html?_r=0>.

wireless and a wired connection is that wireless connection electrical signals are sent as radio waves through the air, while wired connection electrical signals are sent through a physical wire. Bluetooth is a wireless connection that makes messaging over a long distance and rough terrains much more manageable. You can receive and send messages almost instantly. The NXT brick can send wireless electrical connections via Bluetooth as well as several other robot models, such as EV3 and Vex IQ.

Bluetooth-Android Phone Paired with the NXT Brick Controlled Robot

The equipment used for this example will be the LEGO Mindstorms® NXT robot model build, a computer with the software downloaded, and at the Google Play store, download the "NXT Remote Control" app for Android phones. (see reference for google play, macarceller)

 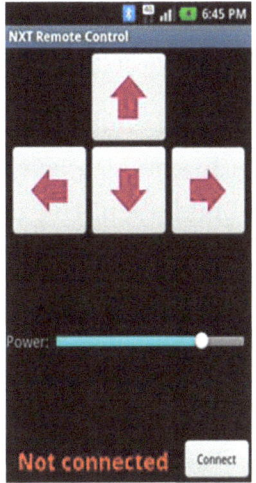 

(Fedorynski, 2016).

Android: A type of operating system designed for smart phones and tablet computers.

General setup procedures will be to turn on the NXT brick, attach the USB cord to the computer, and open up the LEGO Mindstorms software program. Start a new program and name your robot. It is helpful so that you can find your program later. Save your program file for future use.

Once in the software program look for the gray panel of buttons at the bottom right corner of the screen. The left upper right corner of the gray square has an NXT icon brick on it. Click "SCAN." Select the NXT device and press "connect." The name you named the robot should show up (default is "NXT"). Here you can click inside the dialogue box and change the name, also. Close the dialogue box.

To turn on Bluetooth onto the NXT Brick, you will need to use the orange buttons found on the top of your NXT. Scroll to Bluetooth and select "ON." Turn Bluetooth on. On your Android phone, download the NXT Remote Control app at Google Play. Open the app on your phone and click "YES" to accept terms and that it is ok to turn on Bluetooth. At the top of the screen click "Connect" and when prompted, hit SCAN. After a second, the name you gave your robot should appear on the phone. Click the name and you should be connected to the NXT brick. The arrow keys will allow you to remotely control the NXT Robot. Practice navigating as the Android phone sends messages to the NXT brick.

Android App that will control Arduino Robots via Bluetooth

On the web, you should be able to find other apps available for your robot. There is an app for Arduino and also a YouTube.com video explaining how to use the app with your robot, Android app to control Arduino Robots via Bluetooth (mscarceller, 2016).

Android App that will control EV3 Robot via Bluetooth

LEGO Mindstorms Robot Commander for EV3 is a free official command app. Works on several apparatus and attaches by Bluetooth to the EV3 Intelligent Brick. The app is simple to employ and works with your robot. You can build about five different advanced models with the EV3 kit and app.

The Create and Command your robot widget is available in the app for creating your commands with a robot your design. (Commander & LEGO Systems, 2016).[4]

http://itunes.apple.com/us/app/lego-mindstorms-robot-commander/id681786521?mt=8

Private Software Developers

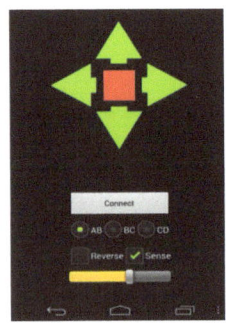

Some applications have been developed by private developers and are available for use. This particular app may not work for all devices even though it states that it will. Last update 12/2014. Once you attach the LEGO Mind storms EV3 brick, buttons will manage the engines. AB, BC, and CD: Select engines Sense Mode: Push and release button handling; Smart Turn with red button Reverse: All engines will be inverted ("EV3 Simple Remote", 2016).

Requirements:

- Tested with Nexus 4 and Nexus 7

- No EV3 program required a previous Bluetooth Pairing is required.

---

[4] http://itunes.apple.com/us/app/lego-mindstorms-robot-commander/id681786521?mt=8

- Bluetooth- Ports A, B, C and D

- LEGO® Mindstorms EV3

- Min. Android 2.1

NXT-NXT Remote Control for Challenge Games

A soccer challenge game can be played with two NXT robots. The wireless (Bluetooth) can be used to connect one NXT robot to control another remotely. NXT "controller" robot and NXT "receiver" robots are paired to work as sender and receiver. You need to gather two NXT Intelligent Bricks, a mini ball and a goal post for one player game.

The "wireless" electrical connection between the two NXT Brick brains allow the devices to link remotely without being tethered. The NXT is already set up with a Bluetooth code block in the software program that allows for the NXT bricks to download the codes into the separate NXT robots. The NXT can send text messages through Bluetooth. (To use Bluetooth with a stationary "autonomous robot," you could build a robot and attach the ultrasonic sensor to detect motion in a dark room. If it detects motion, it could send a message saying "motion detected").

NXT Bluetooth Controller Program

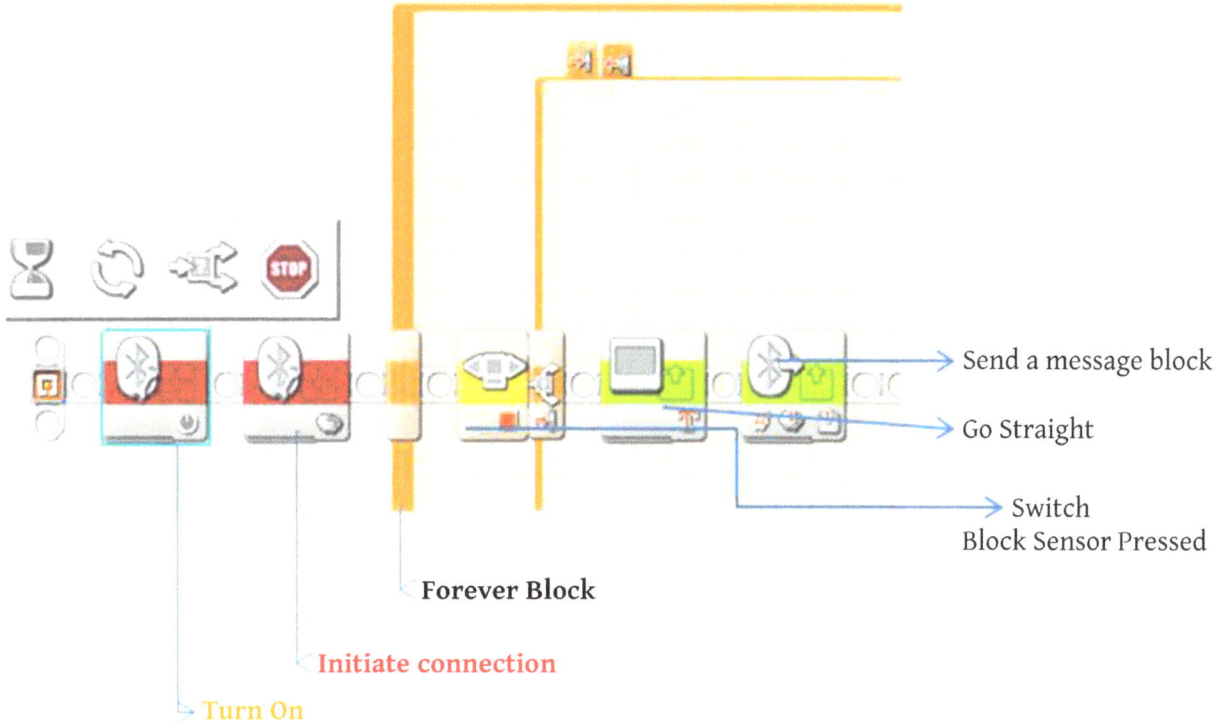

Send a message block

Go Straight

Switch
Block Sensor Pressed

Forever Block

Initiate connection

Turn On

Controller Graphical Program "Algorithm"

- Each button (orange, right, left) is going to be pressed on the controller NXT. It will send a distinct message ("1","2", and "3") via Bluetooth then to the receiver. The receiver picks up the coded algorithm and completes the corresponding tasks listed below.

- When no controller button is pressed on the controller NXT, have the controller send another distinct message ("0") to the receiver.

NXT Receiver "Algorithm"

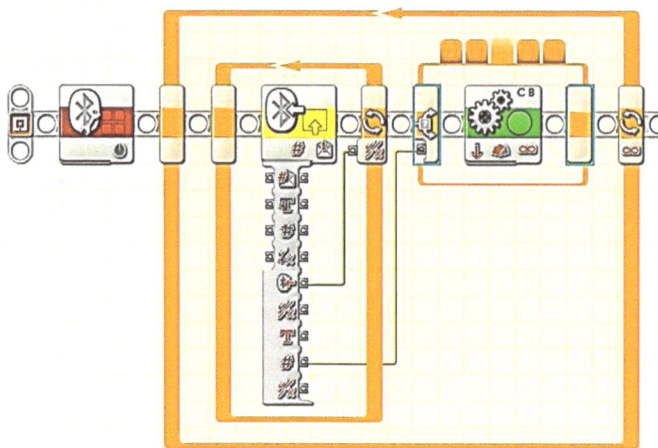

- Program the receiver so that:

- when it receives "1", it moves forward

- when it receives "2", it moves left

- when it receives "3", it moves right

- when it receives "0", it does nothing

Once the receiver NXT receives the "0" message via Bluetooth, it will know that the controller NXT did not have any buttons pushed. It will stop moving. When the receiver gets the "1" signal, it knows that the orange button on the controller is being pushed, and the receiver will move forward.

The receiver receives the programmed number "2", and it knows that the controller's left button is being pushed. Likewise, when the receiver receives the number "3", it knows that the controller's right button is being pushed, and the robot will turn right.

Bluetooth Program

- Both NXT Bricks become the remote control device (controller) and the other robot vehicle responding to the controller is the receiver.

- The controller sends the messages via Bluetooth when you press a different button on the NXT brick brain.

- The other NXT (receiver) gets those messages from Bluetooth, and the robot is mobile moving forward, right, left, or backward when it receives the messages. The other NXT, the receiver, receives those messages via Bluetooth and is programmed to move forward, left, right or backward when it receives the messages.

The two NXT's will work together via a Bluetooth wireless connection.

The buttons and arrows will become your joystick controller. The orange button makes the robot move forward, and the receiver gets the signal. The left arrow and the right arrow will turn the robot in either direction. A ready built robot (receiver) will be able to sync and work together.[5] Practice a field game such as soccer with your Robot and Controller. Try to test it and make improvements.

Make a Simple Remote Controlled Car

There are many great websites to search for building a remote controlled car such as DIYHack.com, Instructables.com, and Pinterest.com; however, I will explain the basic circuitry and simple parts of the car.

Materials for RC Car

1. DC motor

2. Castor wheel

3. Clamps- Motor Mounts

4. Wheels

5. Screws and spacers

6. LEGO® or VEX IQ parts

---

[5] See Resources for YouTube Video by Damien Kee NXT-NXT Tutorials

Electronic Components

Resistor	Motor Driver
Capacitors	Tactile Switch
RF transmitter	Battery Packs
RF receiver	Soldering Iron
5. Encoder/Decoder	

Logic for the RC Robot Car

The robot car works with a logical control system. The **receiver** (robot car) as well as the **transmitter** (remote control).The transmitter has switches to help with digital inputs to the encoder IC, which in turn translates the data sending it to the RF Transmitter module. The Receiver picks up the encoded information and passes it along to the decoder, which in turn decodes the information, sending it along to the motor driver IC to drive the motors.

Flow Diagram

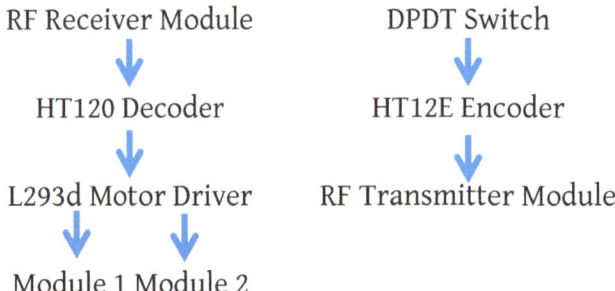

Understanding Datasheets

Manufacturers provide a data sheet or spec sheet document to explain how a component behaves. It will be to your advantage to learn to find the information for the data sheets for the components that you use to build your robot. Google Search the component name, and some links should pop up for the datasheet. Electronic enthusiasts learn to read and understand datasheets so that they can be sure that they know how their components work.

Differential Drive Algorithm

We can look at a basic car to drive. We are using two motors to drive the car. We still use the **differential drive** method, because we want the wheels to turn in more directions than forward and backward. The

castor wheel is a stability wheel that helps to give mechanical stability via some friction and balance to the robot car. It is attached as third wheel to the back of the vehicle. It will look like the castor wheels at the bottom of rolling chairs.

How will the car maneuver around the track if it only has only two wheels? The differential drive algorithm comes into play. The track control is going to be turning both wheels in opposite directions via the Infrared Receiver.

The table below can offer you a better way to understand the turn angles along with forward and backward movements based on the differential drive direction of wheels.

It looks like your car should move in several directions based on this chart. The car goes forward and backward when one pair of motors function in one track while the right and left will make the robot car operate in multiple directions. This is a differential drive.

Left Motor	Right Motor	Direction
Front	Front	Front
Front	Back	Right
Back	Front	Left
Back	Back	Back

RF Transmitter and Receiver Module

With different manufacturers, RF transmitter and receiver modules may look different. The transmitter and receiver should still be the same frequency. Some modules have inverted numbering, but the Pin and Pad should be compatible. More about pins will be found on the data sheets if you get your data sheets for your robot model.

Designing Your RF Transmitter (Remote Control)

The remote will need to be made so that you can handle it with ease. The size is important so that you will be comfortable carrying it around. The transmitter circuit will need to have a good closure for the circuit. The HT12E encoders are 12-bit encoders and it has eight address bits and 4 data bits. The address bits might be left open or pulled low. The address pins (A0 to A7) is attached to a switch if the switched is ON then that line is attached to Ground GND (Vss) otherwise, the pin is left suspended. The transmit enable is an active low input to the **integrated circuit (IC)** and allows the transmission. When the switch attached to pin 14 is pressed, the eight address bits along with the 4 data bits (AD8 to AD11) are sequence encoded and sent out at the Decoder OUT pin. For the application (RC robot car) we will attach the transmit (TE) enable straight to ground GND to maintain forwarding the information to the RC car. The receiver circuit, doesn't require it to have a strong battery. A 9V battery can power that circuit.[6] This is just a short overview of designing your own remote control. Below is a circuit schematics diagram. Check online for a Do It Yourself video.

---

[6] http://embedjournal.com/make-a-rc-robot-car/ retrieved July, 10th, 2016

HT-12E and ht12D Circuit Schematics

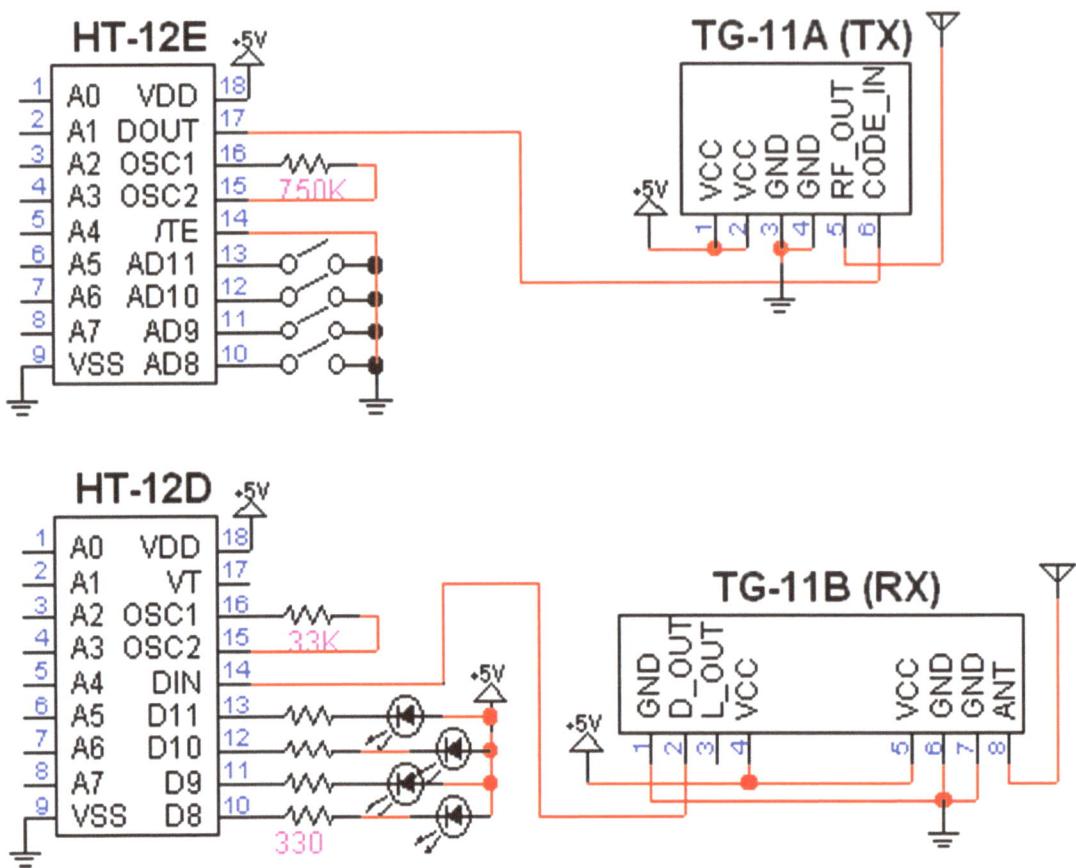

("HT-12E and ht12D Circuit Schematics", 2016).

RF Receiver Circuit with Motor Driver

The RF reception is handled by the receiver. It also has a circuit diagram for the receiver and the motor drive. The encoder and the address pin in the decoder (HT12D) behave the same way. The data is received at the DIN pin (from the RF receiver circuit) and the data is checked three times. This is the reason to study the data sheet; the information is found on the sheet. When the decoder matches the received data from the encoder, the data is decoded and fastens onto the data pins (D8 to D11).

Control signals will be sent to the motor drive IC, and the motor will drive in a forward and backward direction. If you can to make an adjustment to your controller adding a VT (valid transit) pin, add a LED with series resistance that can give off a light to indicate the transmission between the encoder and decoder are working.

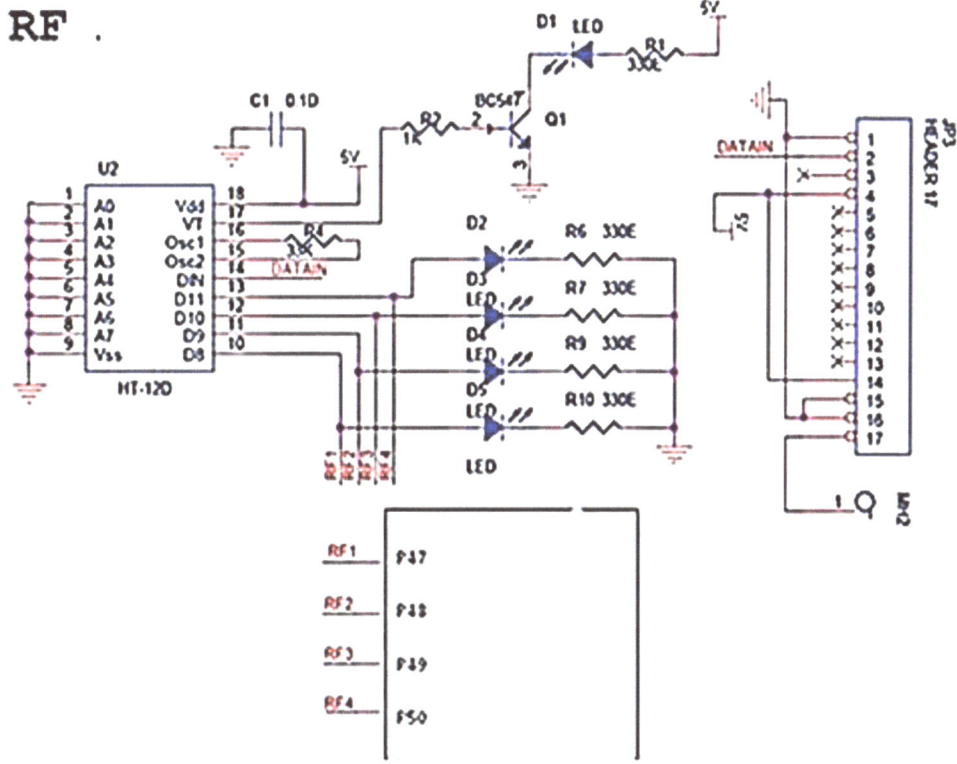

("RF Receiver Circuit with Motor Driver", 2016).

Troubleshooting Your Circuit

If you run into problems with your Remote Controlled robot, retest your RF transmitter and receiver circuit. You might have shorted the circuit due to improper soldering. Power-off the Power Circuit and then test for shorts in the circuit with a multi-meter when it is in connectivity mode. Try to eliminate the short. The Motor Driver Circuit are easy to brick, and most of the time, you should see a crack and sometimes it might smoke. You should also get VCC2 in the multi-meter, if not; it will need to be replaced. The RF Transmitter and Receiver module still do not work; you might need to try another pair and switch them out.

References:

Bluetooth Radio - VEX IQ Forum. (2016). Retrieved 28 December 2016, from
http://www.vexiqforum.com/forum/main-category/vex-iq-product-discussion/10286-bluetooth-radio

File:360 controller.svg - Wikimedia Commons. (2016). Commons.wikimedia.org. Retrieved 28 December 2016, from
https://commons.wikimedia.org/w/index.php?curid=9655347

Turi, J. (2014). Tesla's toy boat: A drone before its time. Engadget. Retrieved 28 December 2016, from
https://www.engadget.com/2014/01/19/nikola-teslas-remote-control-boat/

File: Tesla boat1.jpg - Wikimedia Commons. (2016). Commons.wikimedia.org. Retrieved 28 December 2016, from
https://commons.wikimedia.org/wiki/File:Tesla_boat1.jpg

Fedorynski, J. (2016). Retrieved 28 December 2016, from
https://play.google.com/store/apps/details?id=org.jfedor.nxtremotecontrol

Mscarceller, m. (2016). Retrieved 28 December 2016, from
https://play.google.com/store/apps/details?id=com.mscarceller.makeblockbtcontroller

EV3 Simple Remote. (2016). Retrieved 28 December 2016, from
https://play.google.com/store/apps/details?id=com.EV3.Simple

Commander, L. & LEGO Systems, I. (2016). LEGO® MINDSTORMS® Robot Commander on the App Store. App Store. Retrieved 28 December 2016, from https://itunes.apple.com/us/app/lego-mindstorms-robot-commander/id681786521?mt=8

HT-12E and ht12D Circuit Schematics. (2016). I.stack.imgur.com. Retrieved 28 December 2016, from
http://i.stack.imgur.com/jVNnF.gif

Controller - VEX Robotics. (2016). Vexrobotics.com. Retrieved 28 December 2016, from
http://www.vexrobotics.com/vex-iq-controller.html

Robot Virtual Worlds | ROBOTC | VEX & NXT Simulator. (2016). Robotvirtualworlds.com. Retrieved 28 December 2016, from http://www.robotvirtualworlds.com/download/

Halcli, K. & Frazee, G. (2016). Robot-delivered lethal explosive in Dallas police standoff was a first, experts say. PBS NewsHour. Retrieved 28 December 2016, from http://www.pbs.org/newshour/rundown/robot-delivered-lethal-explosive-in-dallas-police-standoff-was-a-first-experts-say/

Siddharth, S. (2013). Make a Simple RC (Remote Controlled) Robot Car - Embed Journal. Embed Journal. Retrieved 28 December 2016, from http://embedjournal.com/make-a-rc-robot-car/

Chapter 19 Electromagnetic Radiation, (2016)." Physical Science: Teacher Wraparound Edition. New York: Glencoe/MCGraw-Hill, 1999. 529-30 print.

Wi-Fi___33 robots- how to build a Wi-Fi robot

http://www.superdroidrobots.com/shop/custom.aspx/wifi-robots-overview/14/

All On Robots http://www.allonrobots.com/types-of-robots.html

YouTube Videos

Damien Kee, NXT-NXT Controller-Receiver Video Tutorials

Part 1-https://youtu.be/CN3iXGsK9YM

Part 2-https://youtu.be/LlSTK6XSZpw

Part 3-https://youtu.be/LlSTK6XSZpw

Remote Controlled Car with Android phone https://www.youtube.com/watch?v=LuJ7l4kBp-0

CHAPTER 4

Troubleshooting Teamwork and Virtual World Robotics

In this chapter, we are exploring some essentials of working with robots, such as teamwork, troubleshooting, and virtual worlds. If you were to further your education to become an expert Technician in Robotics, your job would be in the field of Automation (Motion Control Robotics). Automation Technician/Engineer education requirements include an extensive mathematical coursework, which includes learning teamwork and troubleshooting. Automation Technicians/Engineers will most likely work in some virtual world setting using AutoCAD and other software programs, which help with prototyping, model design, and mechanics of operating robots.

A college description may look like this one: *Automation, Robotics, and Control are multidisciplinary engineering fields concerned with the design, modeling, analysis, and control of predominantly, computer-based automated systems or processes. Automated systems typically contain a mixture of equipment, devices, software, hardware, and humans. The discipline requires knowledge of elements of electrical engineering, mechanical engineering, chemical engineering, software programming, communication systems, and human factors engineering.*[1] Mathematics is an important element, but understanding the flow of programming, teamwork, and troubleshooting are also essential skills that you will need to learn.

Troubleshooting Robotics

Problems can arise with your robot due to the model design, or the program. Troubleshooting systematically approaches the problem and can help you to find and correct any issues that arise in your robot design, software program, or robot microcontroller. Troubleshooting is a skill that can be sharpened with time, patience, and experience. There are steps to follow that will allow for a quicker solution when you need to find and correct errors with your robot program.

---

[1] http://www.oit.edu/academics/degrees/eere/automation-robotics-and-controls-engineering

Steps can be taken using a **critical thinking map**, also known as a "**Flow Chart**" process. A flow chart will help to gather information on the problem, such as a behavior that is observed, or finding any deficiency in the functionality of the robot that is occurring. Flowcharting can also provide some conditional statements "if-else" process. The "if-else" Statement code may read like this: if (condition) {// true-commands} else {// false-commands }.

The steps to troubleshooting will help to verify the functioning parts in the robot and check all the components. If the problem is identified, it can be repaired, adjusted, or replaced. Evidence of effective troubleshooting results in restored function.

Success often depends on thoroughness and experience of the troubleshooter. Experienced troubleshooters become the "go-to" person who may be dubbed "tech savvy." If you are going to be a trained robotics technician, patience will be a virtue because it may take several attempts and many trials to get your robot working optimally. Websites for high school robotic troubleshooting support can be found online.VEX.com, LEGOengineering.com, and SocietyofRobots.com all have forums to support any troubleshooting issues you have programming your robot or you just want to ask a question.

Flow Charting

The "**flow**" process programmer's use when writing code is an obvious step that gives instructions that are both consecutive and specific. Flowcharts use the steps to work out the commands and conditional statements that translate the behaviors into code. Included in the flow chart are statement blocks and decision blocks creating a Pseudo code that will breakdown the robot behaviors into short commands. Complex behaviors can be explained with simple behaviors line by line, in the order you want the robot task to obey, or describe the action needed to be taken. Move Forward and Stop, etc. Example: The ultrasonic sensor is prompted to stop if it senses something at a distance less than 20 cm. It is programmed to stop. You can further code it to back up if it senses an object.

For each complex behavior, break it down into Simple Behaviors line by line in the order that each should happen. (See the chart below).Try to describe actions and what prompts each action.[2]

Below is a pseudo code outline of an if-else Statement Boolean Logic.

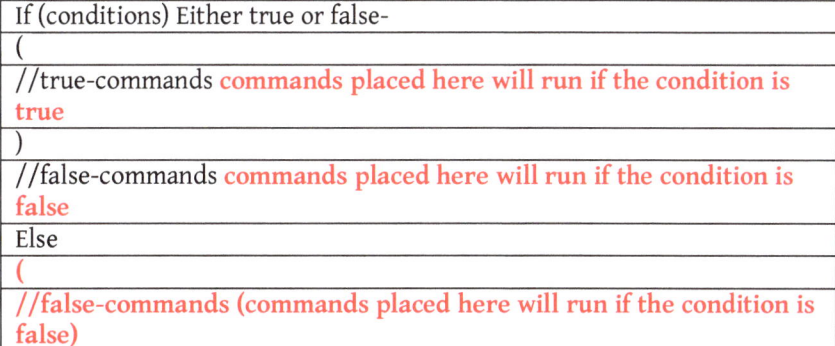

If (conditions) Either true or false-
(
//true-commands commands placed here will run if the condition is true
)
//false-commands commands placed here will run if the condition is false
Else
(
//false-commands (commands placed here will run if the condition is false)

---

[2] ("Pseudocode Programs in RobotC", 2016).

The condition can be true or false as in Boolean Logic. **Boolean algebra** is the branch of **algebra** in which the values of the variables are the truth values *true and false*, usually denoted 1 and 0 respectively.

task main()
(
{ while(true) –
(
{ if(Sensor Value (Ultrasonic Sensor)>25**)** **(25 represents the distance the sensor can detect an object in Centimeters)**
{ motor[motorC]=100;
motor[motorB]=100; **}** **(both motors run full speed if the commands are true)**
)
else
(
motor[motorC]=0;
motor[motorB]=0; (robot will stop moving the sensor has detected less than 25 centimeters-an object may be in the way)

An "**If-Else Statement**" is one way you allow a computer to make a decision. With this command, the program will check the condition and then execute one of two pieces of code, depending on whether the condition is true or false.[3]

Scenario: You are a VEX IQ "tech savvy" technician. The person everyone goes to when their robot is not functioning well. Your job is to troubleshoot the robot's issues. You will be working together in a team to solve the problem based on the process of elimination. There are software and hardware issues, like slow performance, incorrect output, and error messages. You can apply a flow chart that helps to keep the troubleshooting timeframe down. A flow chart is a systematic approach to troubleshooting also known as "**continuous improvement**."

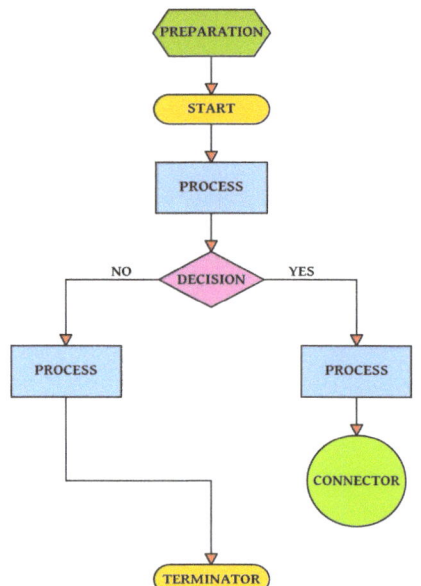

A good troubleshooter will be able to separate, clarify, and prioritize the symptoms in a practical way based on critical thinking, experience, and using the process of elimination flow chart. He/she can usually test the causes and determine the corrective measures.

---

[3] http://cdn.robotc.net/pdfs/natural-language/hp_if_else.pdf

Problems can arise in the **end effectors** (mechanical arm attachments) or in the gripper, pneumatic system, and software program. Most likely, the software program will need to be corrected and tested several times to get your robot to operate efficiently.

There should be a "check sheet" to document the data of your trial and error troubleshooting process. A check sheet is just a simple form or document that can be filled out so that others involved in your team can quickly evaluate what you have already tried to fix. You will not want too much downtime because that causes unproductive blocks of time. In an event such as a **FIRST®** (*For Inspiration and Recognition of Science and Technology*) competition, downtime can cost you to the competition.

Another troubleshooting tool you can use is called a **fishbone diagram**. It is a visual diagram noting causes and effects of the robot's problems and drawing conclusions about teamwork performance. This type of diagram is designed to note causes that are added like the branches of a fishbone structure. Then as a team effort, causes are analyzed and compared to effects, good or bad during a competition. An example of the fishbone applied to a FIRST® Team competition is detailed by Christopher Fultz, (2017) Mr. Fultz is a Robotics Competition Cyber Blue 234Team Mentor from Indiana.[4] He outlines importance of the fishbone diagram and the importance of learning and improving from your mistakes. Mr. Fultz contributes an excellent troubleshooting guide based on his own teamwork experience in FRC Cyber Blue 234 Teamwork competitions. The following Fishbone diagrams explanations used by the Cyber Blue 234 Team provides a clear and concise troubleshooting methodology.

How to Learn and Improve From Your Mistakes

As businesses grow and evolve, they learn from their shortcomings and work to continually improve their performance. This process of continuous improvement is why products continue to improve in performance, quality, and features and often can do so at a reduced price.

FIRST Robotics teams can also benefit from continuous improvement processes by formally evaluating past performances and identifying ways to make improvements.

Continuous improvement comes from taking a good look at results and identifying what has been done well, and what can be improved. The goal is to learn from mistakes and avoid repeating them. An effective continuous improvement program can help you prevent the need to say "That always happens" or "We always end up like that."

One method of a Continuous Improvement process is called **Root Cause** and **Corrective Action** or "RCCA." In an RCCA process, the undesired result, or outcome, is examined to identify all of the contributing factors to the specific result obtained. These contributing factors are then further investigated to identify root causes. Actions are then put in place to eliminate the undesirable root causes.

An effective way to begin an RCCA process is to get a team together and brainstorm ideas regarding the undesired results. One structured brainstorming process that is widely used is

---

[4] https://www.chiefdelphi.com/media/papers/tags/teambuilding

'Ishikawa' diagramming. (The diagram concept was developed by Kaito Ishikawa in 1943 while he was at Tokyo University). When complete, these diagrams resemble fish bones, and they are often referred to as fishbone diagrams.

The initial fishbone diagram looks like this:

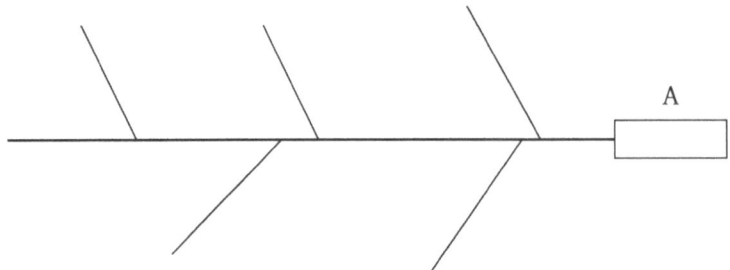

"Courtesy Chris Fultz and FRC Team 234"

The process: There are six basic steps to follow:

1. Agree on the 'undesired result' to be evaluated and draw the fishbone diagram.

2. Identify the main legs of the fishbone.

3. Brainstorm ideas that could cause the undesired result and list them by what leg they apply.

4. Review and identify the most relevant factors or groups.

5. Identify specific actions and action plans to address the main factors.

6. Complete the actions and see the benefits of your work.

1. The fish head (A), is where the undesired result is listed. For a FIRST team, this could be poor performance in a match, a robot function that did not work properly, or a team activity that did not go well.

2. After the undesired result (A) is listed, each of the main legs of the diagram is labeled. These legs (B) are to identify major categories of issues that may be contributors to the undesired outcome. In many manufacturing processes, these legs are Material, Measurement, Machine, People, Methods (Processes), Timing, Resources, and Communication.

After adding these main legs (B), the diagram now looks like this:

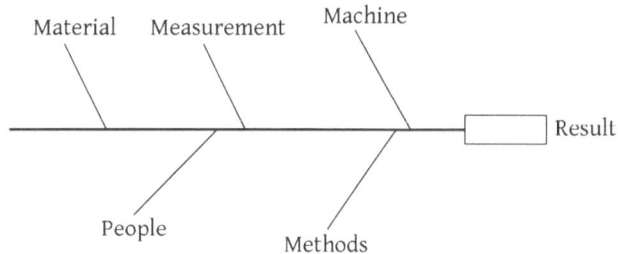

"Courtesy Chris Fultz and FRC Team 234"

3. Next, the team will brainstorm ideas that could have had an effect that caused the undesired outcome. These ideas are grouped in the major categories and written as smaller 'bones' on the main legs. Often, one idea might be listed on more than one leg. The team continues to brainstorm ideas until nothing new is being thought of.

Be Candid

It is very important to have open, candid discussions during this phase. Participants must be sensitive to others, but at the same time open to constructive criticisms about past performances.

After adding more ideas, the diagram now looks like this: (imagine that each line is an idea on that leg).

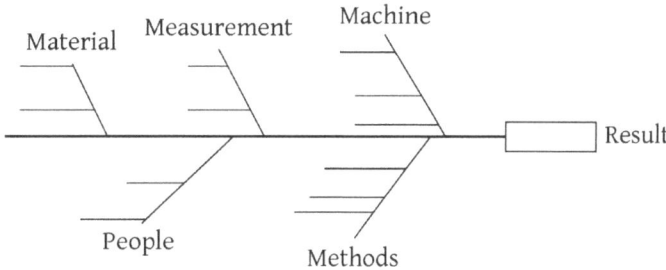

"Courtesy Chris Fultz and FRC Team 234"

4. The next step in the process is to do a quick review and identify items that appear to have the most significant impact on the undesired outcome. Action plans will need to be generated for these items.

5. This step is to identify specific actions and action plans that will eliminate the identified issue and then improve the team performance. (Simply identifying issues and determining a cause will not make performance any better the next time. The path to continuous improvement is to create and implement action plans.) Actions must:

a. Be Specific—clearly describe what must be done

b. Be Assigned—Assigned to a specific person when possible. One accountable person is much better than assigning an action to an entire group or team

    c.   Have a due date—so the person knows when to be complete

    d.   Be measurable—so that it is clear when the action is complete

    e.   Be tracked—so that status and any problems can be identified.

6.   The next step is probably the most challenging – and that is the completion of the action items. All of the effort is mostly wasted if the identified action items are not pursued.

It is a good idea to complete a "ranking" process to determine which actions should be completed first. Usually, there will not be enough time or resources to complete every action, so a ranking process can help set priorities of what actions to do first.

An easy method is to classify each action in one block of a four-block diagram.

The Y (vertical) axis is "impact," or how big of an influence completing the action could be for correcting the undesired results. Low impact items are low on the scale; high impact items are high on the scale.

The X (horizontal) axis is "cost." The cost could be in dollars, or time, or weight. Low-cost items are to the left; higher cost items are to the right. (The scales do not need specific values assigned—relative impact and relative cost are fine.)

Items in the number 1 box (low cost, high impact) would be the first to get tried. A team might then choose some number 2 and some number 3 items to work – this would give a mix of high cost but high impact items with some low cost and low impact items. Usually, the items in box four would not be worked on until everything else was finished, or they might never be tried since they are high cost and low impact.

This diagram looks like this:

HIGH	1	2		1 = Items that are high impact and low cost to implement.
IMPACT				2 = Items that are high impact, but more expensive to implement.
LOW	3	4		3 = Items that are low cost, but also low impact
	LOW <COST> HIGH			4 = Items that are high cost, but low impact.

The six basic steps that were followed:

1. Draw the fishbone and agree on the 'undesired result.'

2. Identify the main legs of the fishbone.

3. Brainstorm ideas that could cause the undesired result and list them by what leg they apply to.

4. Review and identify the most important factors or groups.

5. Identify specific actions and action plans to address the main factors. Action items should:

    a. Be Specific

    b. Be Assigned to a specific person

    c. Have a due date

    d. Be Measurable

    e. Be Tracked

6. Complete the actions and see the benefits of your work.

    a. Actions can be prioritized so that the most important ones are worked first.

Example: We will use an example from FIRST Robotics.

Let's assume that team 9999 had a good season, but its robot had a few problems during competition. One of the problems was that electrical connectors came loose in about one-quarter of the team's matches. Sometimes it was not a big problem, and they were able to play well, but other times it made them lose their matches. The team would like to understand all of the potential factors that could cause their undesired result (loose electrical connectors) and find a way to perform better next season.

1. Sketch the fishbone diagram and identify the undesired result of "Electrical Connectors Loose."

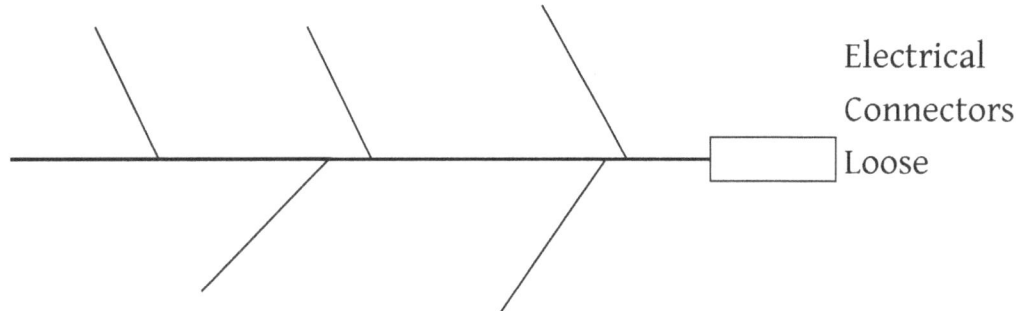

Electrical Connectors Loose

"Courtesy Chris Fultz and FRC Team 234"

2. Identify the main leg of the fishbone. We will use the five common legs.

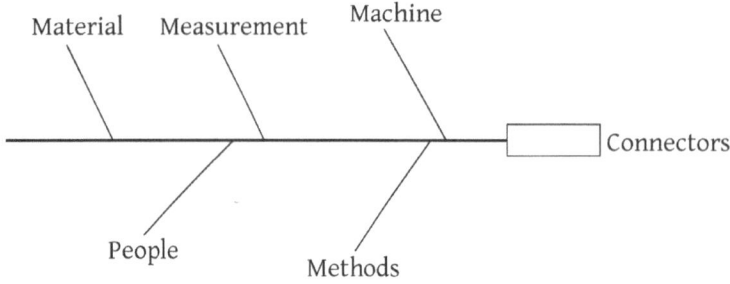

"Courtesy Chris Fultz and FRC Team 234"

3. The team will brainstorm ideas and list them on the correct leg.

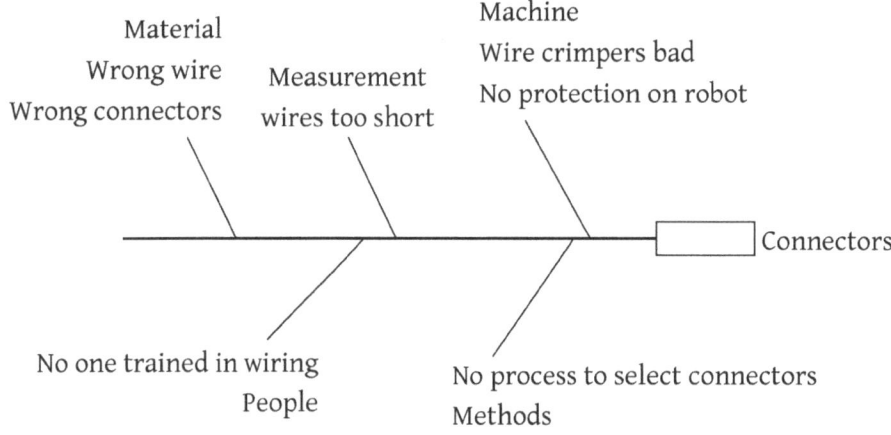

"Courtesy Chris Fultz and FRC Team 234"

4 & 5.   The team will now list each of the items and try to create action items to address the identified issues.

Main Factors:

1. Wrong wire was used

2. Wrong connectors were used

3. Wires were too short

4. Wire crimpers were bad

5. There was no protection on the robot

6. No one was trained on how to do wiring

7. There was no process to select the right connector

Actions:

1. We will make a chart of wire sizes and the correct connectors.

2. We will give each wire at least a 1" slack when installing.

3. We will have two people trained in the correct process to crimp wires.

4. We will buy a nice set of crimpers.

5. We will recruit an electronics person to teach us to wire correctly.

   The team would then assign each of these to a specific person or small team, define a due date, determine how they will know when the action is complete, and request regular status reports on longer term actions.

   Actions might be ranked by priority, so the most important items are worked first.

6. Complete the actions.

   To be effective long-term, the action and effects need to be tracked. For this team, they would monitor how the robot competed during the next season. If the team had implemented all five actions and then had no loose connectors during their next season, they could assume they had found the cause, and corrected it. The team could then move on to a new issue and work to eliminate the cause of it.

If, however, the team continued to have problems with connectors the next season, they would need to recheck their work and determine what they had missed and continue their efforts to find the true cause.

Summary

"Continuous Improvement activities can help a team improve its performance each year. This performance improvement might be on the competition field, in recruiting and maintaining sponsors, in community service or many other areas. Root Cause and Corrective Action analysis, through fishbone diagramming, is a straightforward process that teams can use to help identify areas for improvement.

By identifying areas that need improvement, and then creating specific actions and action plans, teams can continue to raise their level of performance from season to season. Since these principles are widely used in industry, students that learn and use the basics of Continuous Improvement will have an advantage as they move through engineering studies and into the workforce."

Continuous Improvement

James Crowder Engineer Raytheon Company

Permission to Use granted by Mike Doble, Rayetheon 06-01-2016

Artificial intelligent robots have been of interest to engineers for many years. Engineer James Crowder with Raytheon. Raytheon Company is a technology and innovation leader specializing in defense, civil government, and cyber security solutions. Founded in 1922, Raytheon provides state-of-the-art electronics, mission systems integration, capabilities in C5I (command, control, communications, computing, cyber, and intelligence), sensing, effects and mission support services. Raytheon, headquartered in Waltham, Massachusetts, has built two robots that Crowder designed to improve continuously as part of his research in developing robots that have some system of reasoning and learning autonomously. In the context of the ongoing continuous work, he seeks to improve Zeus and Hercules (his artificially intelligent robots). Crowder wanted to start with something small before tackling a big project like the C-3PO.

Zeus and Hercules are robots with simulated neural systems similar to those of cockroaches and octopuses. Crowder is continuing to improve their development with greater complexity and robotic capabilities from these small creatures like prototypes to eventually build larger more sophisticated robots. The Zeus "robot" cockroach is an amazing little model designed with some simple programming features that make it artificially intelligent Just as a real cockroach detests light, so does Zeus. It can detect light and move away from it by turning or backing up. With only a few codes it was learning to walk and seek darkness. Crowder works with simulating emotions, including simulating thinking and reasoning using mechanical versions of insects and octopuses. Crowder is enabling his fellow engineer peers and himself, to replicate human thought to eventually design human-like robots that can think more clearly, and do things faster, better and smarter.

Virtual Worlds- 3D Modeling

Would you like to practice building and programming a robot without the physical equipment? It is possible with high-end simulation environments such as Robot Virtual Worlds (RVW). RVW is an online program used in schools. However, there is a ten-day trial that you can use to practice using robots

("Robot Virtual Worlds | ROBOTC | VEX & NXT Simulator", 2016). RWV simulates some real world robotics that will give you some hands-on experience using the popular robotic kits such as Vex® LEGO® and TEXRIX®. The environment is 3D and uses ROBOTC to program in the virtual world. Another good resource: ("FIRE Website : Virtual Worlds", 2016).

ROBOTC language simulator is available to practice at ("RoboCatz FLL First Lego League Team #144", 2016) instead of a physical robot. There are two missions you can complete that show the RobotC code and actual simulation in real time. ("Robot Simulator: Teaching JavaScript Programming", 2016).

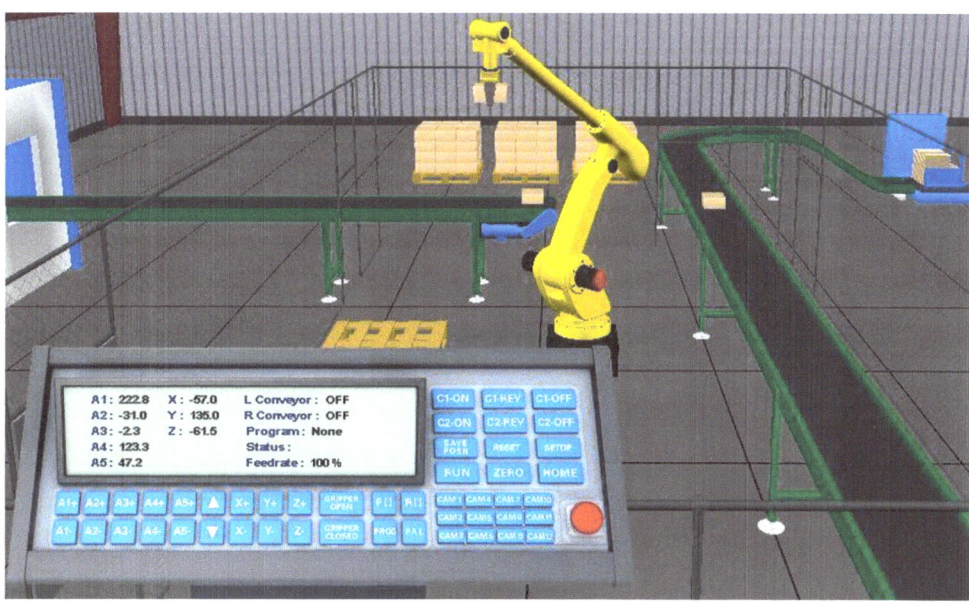

Virtual World Simulator("RoboLogix", 2016).

It can be difficult to learn to program if the lack of equipment and tools affect your motivation to continue learning how to program. Virtual Reality (VR) frameworks allow students to receive some interactive programming instruction within an engineering education course. Virtual Reality is going to be helpful to graduates in the real world of engineering, because engineers use VR platforms in the design process and programming, before building the final prototype.

Virtual Worlds (VW) allow for collaboration and teamwork as an ongoing, continuous improvement in the world of technology. The advantage of the VW is that it supports the teams, allows for reviews, reduces the time for development, and reduces overall costs for the engineering firm the military that has used 3D immersive prototype models to design and control facilities for submarines.

Alice- Learn Computer Programming Animation

Alice can work with NXT bricks. You can use a robot along with the animations. Alice is a free downloadable programming environment that teaches some basic programming in an animated 3D virtual world setting. You can create games and perform tasks with Alice. Alice will work with the NXT brick. However, the firmware within the NXT brick needs to be changed out before downloading the Alice NXT program into the brick brain has the file available for download ("Storytelling Alice", 2016).

Storytelling Alice was created by Caitlin Kelleher, as part of her doctoral work in Computer Science at Carnegie Mellon University. Ms. Kelleher designed Storytelling Alice so that students could learn to create computer programs by creating short 3D animated movies. Storytelling Alice enables users to make animations by building a story with social interactions between characters, and a gallery of 3D characters and scenery with custom animations. Gaining some hands-on programming experience with Alice will help you to understand some fundamental concepts to learn to write code.

References

Pseudocode programs in RobotC. (2016). Robocatz.com. Retrieved 30 December 2016, from
http://www.robocatz.com/pseudocode.htm

Robot Virtual Worlds | ROBOTC | VEX & NXT Simulator. (2016). Robotvirtualworlds.com. Retrieved 30 December 2016, from http://www.robotvirtualworlds.com/

FIRE Website :: Virtual Worlds. (2016). Education.rec.ri.cmu.edu. Retrieved 30 December 2016, from http://www.education.rec.ri.cmu.edu/fire/virtual-worlds/

RoboCatz FLL First Lego League Team #144. (2016). Robocatz.com. Retrieved 30 December 2016, from http://www.robocatz.com/

Robot Simulator: Teaching JavaScript Programming. (2016). Robocatz.com. Retrieved 30 December 2016, from http://robocatz.com/simulation-launcher.htm

RoboLogix. (2016). En.wikipedia.org. Retrieved 30 December 2016, from https://en.wikipedia.org/wiki/RoboLogix

CHAPTER 5

What Can Robots Do?

As we begin this chapter, let's take a look a few things robots may never be able to do. Why? It is important to understand that robots can never replace all jobs in manufacturing, military, and society. The robot is programmed only to be as smart as the programmer. Author, Ian Pearson, a fellow at World Academy of Arts and Science, narrows down three categories of human jobs that robots will never be able to replace. As Pearson explains; "First, teachers are irreplaceable because robots could never relate to, or understand kids. Second, even though defense jobs are becoming increasingly reliant on robots (think drones in the military). Pearson doesn't foresee robots replacing Police and Security in the near future. Third, people who work in management, especially jobs that entail personal or motivational leadership, have security." Do you think a robot would make a good football coach? That might not work for team morale. There are other things robots cannot replace should give you some idea of job security.

Mechanics and Applications of Robots

We have surveyed the inside of a robot. Now we can begin to put parts together to make them move. A robot built from Arduino, or kits like Vex 1Q and EV 3 LEGO® can be built in similar ways to complete tasks. Newer robot kits like the VEX IQ have a Gyro Sensor and can give a robot added advantages with mobility. In this chapter, we will explore mechanical features of robots, apply some math, and write the code for some robots. We will learn what the robots can do, how they can move, and some practical uses for them in the physical world with real life applications.

The Essentials-Programming

The basics of programming begin with understanding **behaviors** that make the robot move, such as straight, turn, or move forward. The basics of programming are also known as Algorithms. One of the first challenges a student might complete would be the Maze Challenge. The challenge is to program the robot to move forward and steer using the software programs graphic interface. The codes for the behaviors are programmed based on the "problem" and how you want to create a plan for the robot to follow when it completes the steps or tasks. These are a sequence of behaviors, and in the table below is a breakdown of what the behaviors would look like in a **pseudocode** translated into the software program. A "pseudocode" is a code written in English that the robot can understand. There are three levels and categories of behaviors, basic, simple and complex. As you can see the behaviors build upon the codes that the robot can understand. The behaviors will have similar behaviors, however, depending on the level of behavior, the robot moves as a basic task of turning the left motor on at half power. The simple behavior makes the robot go forward 2 seconds at half power, and the complex behavior moves the robot around a corner. These behaviors all look similar, but as the code is tweaked, the robot will accomplish higher order of commands.

Main Task			

motor[leftMotor] = 63;
 motor[rightMotor] = 63;
 wait1Msec(2000);
	Basic Behavior		

motor[leftMotor] = —63;
 motor[rightMotor] = 63;
 wait1Msec(400);
	Basic Behavior	Simple Behavior	

motor[leftMotor] = 63;
 motor[rightMotor] = 63;
 wait1Msec(2000);
	Basic Behavior	Simple Behavior	Complex Behavior
			 Combines all Behaviors

Advanced Programming

Advanced programming geared to get the robot moving uses switches, multitasking data wires, display data, operations, and variables. The Ultra Sonic Sensor will have codes that read **"forward until near"**, using sensory feedback messages. The programming works with the actuators, the "muscles" of the robot. Actuators convert stored energy to make your robot mobile. The most well-liked actuators are electric direct current motors that rotate a wheel or gear, or linear actuators. They can be used with industrial robots in factories. The actuators are not only the "things" that make your robot move, but allows it to interact physically with the environment. Manipulators are end effectors that usually grip and lift objects.

Basic Robot Design

The basic robot has two wheels with a caster wheels in the back. The castor wheel helps to balance with a small amount of friction that drags the robot from behind. Geometrically, it is the third point on a plane. It drives under the most basic algorithm. An algorithm is a step by step set of operations that will be performed and is written as a flow chart. Each algorithm can be coded so that the drive-straight code makes both wheels move forward and backward at the same speed, turning left, or right as the individual wheels move in reverse while the other moves forward. The basic robot is designed with a differential drive, two motors, and two wheels.

Real World Applications

Factory Robot

Imagine that you have a client in need of a companion robot for a factory. The robot is powered by a microcontroller, programmed to seek out certain colors and do some sorting. The client is an industrial manufacturer who needs a robot that can sense (pick up) certain colors to sort the stems from the strawberries. The sensors will be set to recognize red, detect and reject leaves, snails, unripe, molded, and rotted discolored strawberries. Amazing? This machine is already in existence with The Manufacturing Company TOMRA. In 2011, TOMRA acquired Odenberg, a company possessing unique, patented technology and leading market positions in several fast-growing segments of the food sorting and processing industry. Complementing this further with the acquisition of BEST, another sorting company, made TOMRA one of the world's leading food sorters, and an expanded technology portfolio unrivaled by competitors.[1]

Color Sorter

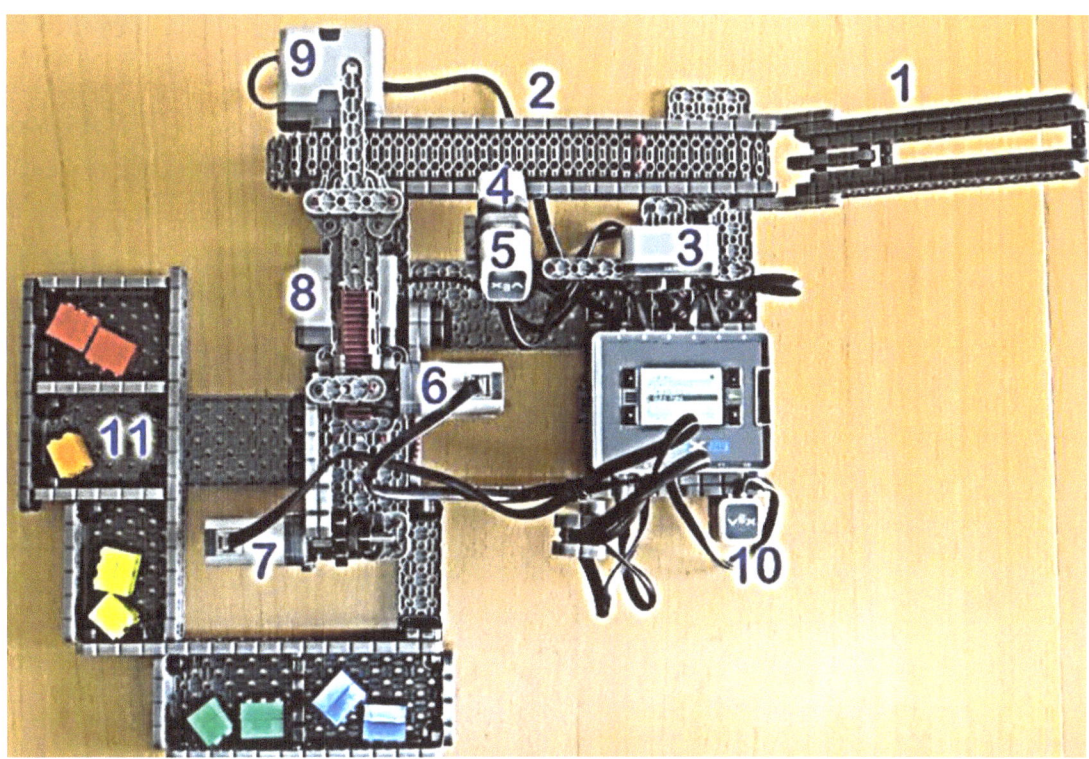

("Philo's Home Page", 2016).

---

[1] Titech to Expand Sensor Based Sorting Range with ... (07/18/2016)). Retrieved from https://waste-management-world.com/a/titech-to-expand-sensor-based-sorting-range

In this example using VEX IQ robotics, the blocks represent the different sorting tasks the machine will be able to complete.

Sorting

1. Block chute. Blocks to be sorted are placed here.

2. Conveyor belt. It picks the blocks from the chute and brings them to the distance sensor, and then color sensor and picker arm. Two pins on the belt catch the blocks.

3. Distance sensor. Detects blocks as they travel on the conveyor. If no block is detected after a half belt rotation, the ramp stop, the Touch LED 10 blinks and waits for the press to start again.

4. Color sensor, activated when the block detected on three is in front.

5. Touch LED, displaying detected color.

6. Gripper motor

7. Arm lift motor

8. Arm rotation motor

9. Conveyor belt motor

10. Sorting restart

11. Output bins

A sorting program will look similar to this program. The parameters are set with a repeat, and the code will wait if/else is the conditions programmed into the code. The color green is the stem that will be removed while the strawberries travel down the conveyor belt.

Graphical Interface for a Color Sorter

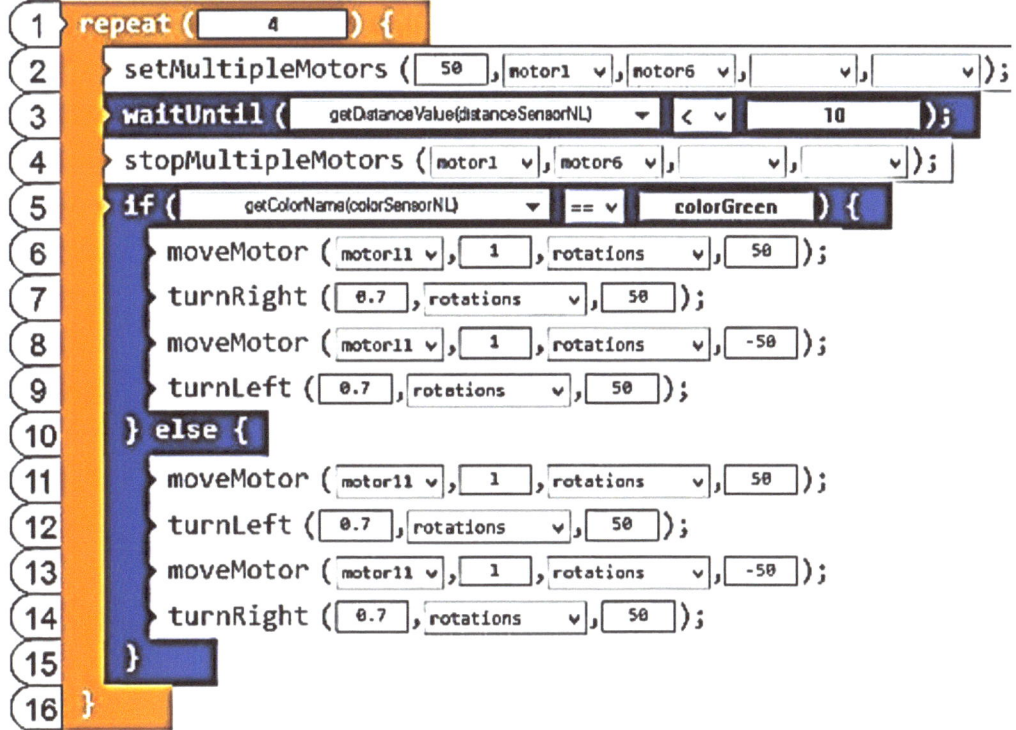

Warehouse

Imagine you are an Engineer and work for a large warehouse. You are designing robots that can move goods along a complicated path traveling up and down ramps, obstacles, and over different kinds of surfaces. They will need safety features built in for factory workers and a navigational system to keep from getting lost. You will need to design and build a robot that easily and safely navigates from one room to another, using pathways (lines) that the robot can follow. Robots programmed to speed up the shipping process for Amazon are based on this same type of warehouse environment. They are automated robots that work in their warehouse running up and down the aisles of the warehouse. Amazon ships millions of packages a day, 24 hours a day and buyers can get packages in two days thanks to these handy robots and mechanical arms.[2]

Line Tracking Robots are used for Automated Material Transport Systems. They move parts around factories by following lines on the ground. The program is written to keep the robot on the lines. As the robot navigates through the factory, it is repeating a conditional behavior.

Program Your Robot to Follow a Line

Programming robots to follow lines is one of the most useful applications in industry because robots are programmed to move from one point to another. In this example, the base of the robot is configured with

---

[2] Meet Amazon's New Robot Army Shipping Your Product." YouTube. N.p., 14 Dec. 2014. Web. 20 May 2016.
https://www.youtube.com/watch?v=g6DIFpaoI6A

the color sensor attached and is wired to port 3 (see graphic program below). The repeat block is (forever). A conditional loop block is programmed so that the robot drives toward the edge of the line where the white floor meets black line. There is a black tape on the floor, and your robot needs to read it. What the robot sensor is reading, "See" black go left and forward, "see" white go right and forward. It is not following the line, but following where it is 50 percent black and 50 percent white from the tape to the floor.

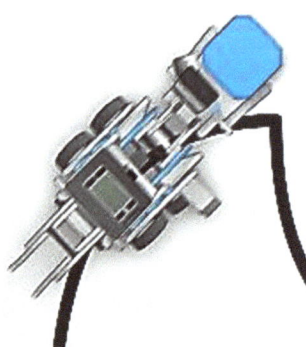

The Sensor will be facing down. The basic code in the graphical interface, like RobotC, would look like this drag and drop the program.

```
1  repeat ( forever ) {
2      lineTrackLeft ( port3 ▾ , 50 , 50 , 0 );
3  }
4
```

EV3 Graphic Interface

EV 3 software interface would look like this. It is a different software programming environment, and the use of pictures makes it easy to learn to program.

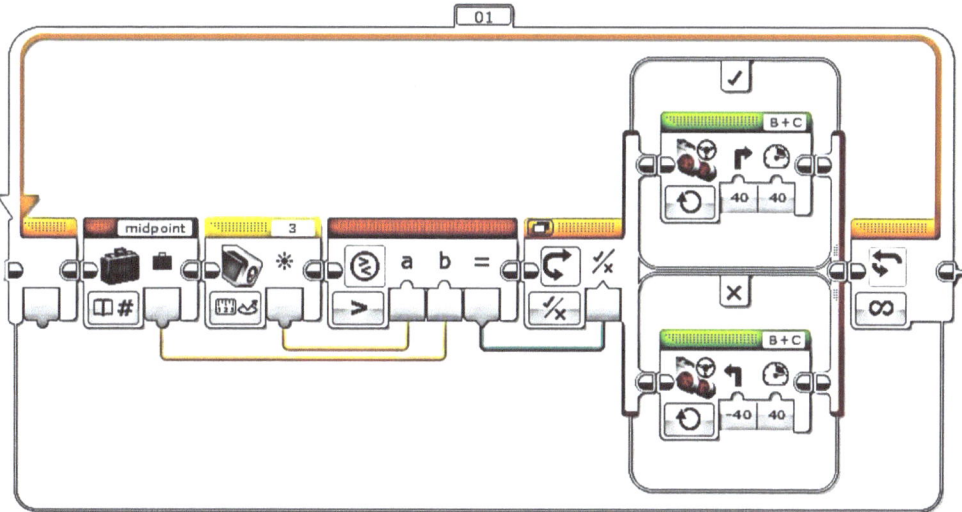

The line tracking sensor has a Threshold Value. It is calibrated before beginning to start the robot on the black line. The instructions are available in the kits. Make sure that your sensor is not too far from the floor as some sensors read ambient light and will perform better if they are closer to the colors it is supposed to read (black for your line follower robot). You should be able to place a penny under the sensor while it is on the floor to gauge the distance of the sensor to the floor. We use this method with the NXT and EV3 sensors. To calibrate, hold the sensor over the black line. It will show a numerical value (in the brick brain). Record the average value of the numbers (the one that is displayed the most). Thenhold the sensor over the white line, it should show white value. Vex IQ comes with an out-of-the-box default sensor already programmed; some other sensors will need to be calibrated.

To calculate the most accurate threshold value for your environment, add the two values that you recorded and divide by two. Example:

Value on Black line = 10
Value on White floor = 90

Threshold equals (10 + 90) / 2 = 50

This number is recorded in the block code above. The last two blocks refer to the motor speed. When the robot is coded to track the line, the color sensor will read in greyscale mode. The amount of reflected light under it has specific ambient features. **Dark surfaces absorb light, and bright surfaces reflect it.**

The Articulated Arm

Imagine that you are an aerospace engineer in the space program. Design, build and program a robot that is built onto a space rover. It will be able to pick up rock samples of different shapes and sizes, and differentiated levels of terrain. The articulated arm will need to be programmed to be controlled easily via a remote controller.[3]

---

[3] VEX IQ Robotics Camp Handbook - manualzz.com. (06/28/2016). Retrieved from
http://manualzz.com/doc/7474775/vex-iq-robotics-camp-handbook

This robot may have linear actuators that move in and out of pneumatic and hydraulic oil systems. All devices that are at the end of the arm are used for "tooling." The end of the arm (grippers and end effectors) tools make your robot interact physically with the environment. A factory built articulated arm design, would be custom designed and built for your specific application for increased speed and accuracy.[4]

Principle	Kinematic Structure
Cartesian Robot	
Cylindrical Robot	
Soherical Robot	
SCARA Robot	
Articulated Robot	

[4] Titech to Expand Sensor Based Sorting Range with (06/27/2016). Retrieved from https://waste-management-world.com/a/titech-to-expand-sensor-based-sorting-range

Vex IQ Arm Bot

This arm is stationary and can pick up objects, bend and pivot around to place the object in another area. A good example of what an engineer would design for the articulated arm. It has a gripper attachment and three sensors attached, and it is stationary.

Programming- You can use a handheld remote control that has a program already written into the default program (already programmed) to carry out specific instructions. The movable parts are the shoulder, elbow, and claw. The Driver Control program in (2) joysticks mode will allow you to pivot the base left to right, the elbow and should move up and down. Now, put together this simple ArmBot build, and we can relate it to how it works in a real environment.

("Rambo IQ", 2013). /Flicker Attribution-No Derivs 2.0 Generic Vex IQ [5]

City Navigator

Imagine that you are an engineer assisting in the development of an automated transportation system. The streets are busy and your automated vehicle (car) will need to have sensors and be programmed to

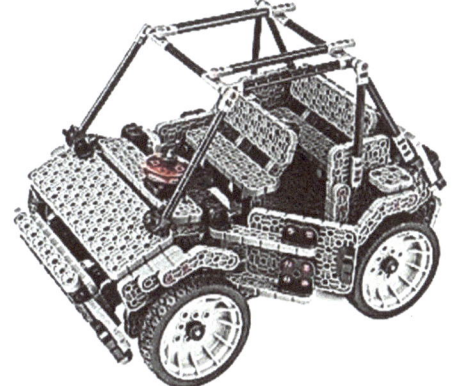

take passengers through a maze of twists and turns through the hustle and bustle of a busy downtown street.

The VEX IQ® calls this demo car a Smart Car. It has sensors that allow for autonomous and hybrid control of VEX IQ robots and other creative model builds. The sensors connect to the robot, and mechanisms are easily programmed. You will need to program the time and distance, rotation, and allow for sense of touch, since you do not want your autonomous vehicle to run into the back of someone. Just what an engineer might need to design for this city transportation center.

Copyright - ©2014 Ken Stanek/kenstanek.com Original Transmission Reference – 20140426
Copyright Notice - ©2014 Ken Stanek/kenstanek.com

Search and Rescue

Search and Rescue robots have become a valuable commodity that can have many advantages for rescue workers, and save more lives than a mere human could do on their strengths. They can locate people, clear routes, measure the temperature in an environment, climb stairs and operate vehicles. One new design can hear voices and alert rescue workers to the location. Imagine you are heading up a team of

---

[5] https://creativecommons.org/licenses/by-nd/2.0/

engineers to design and build a robot that can complete a few search and rescue tasks. First, you will need to collaborate with your team to decide on a Flow Chart System.[6]

The Flow Chart will help the robot think logically and make decisions and act upon them. It is a decision-making diagram, a graphical representation of the robot's plan of action. It is important because the engineer that designs the robot works collaboratively with the programmer, to a quality product for the company. The program will use different blocks that will make the decisions, like loops, switch block, and sensors.

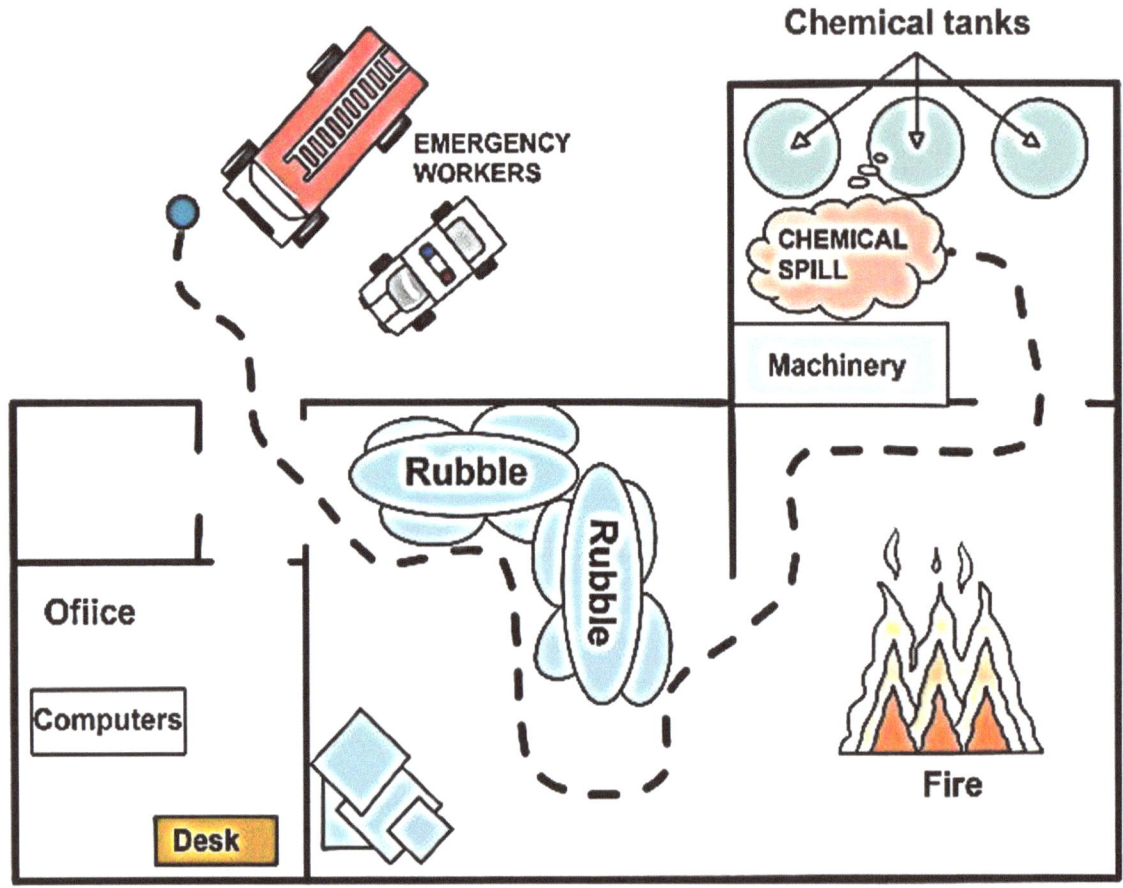

---

[6] Flowcharts." Springer Reference (2006): 1-3. 4-h CCS. National Robotics Engineering Center, 2006. Web. 22 May 2016. http://www.education.rec.ri.cmu.edu/robots/4H/roboticsandyouonline/explorer_content/explorer_7/flowcharts.pdf

Pseudocode for Search and Rescue

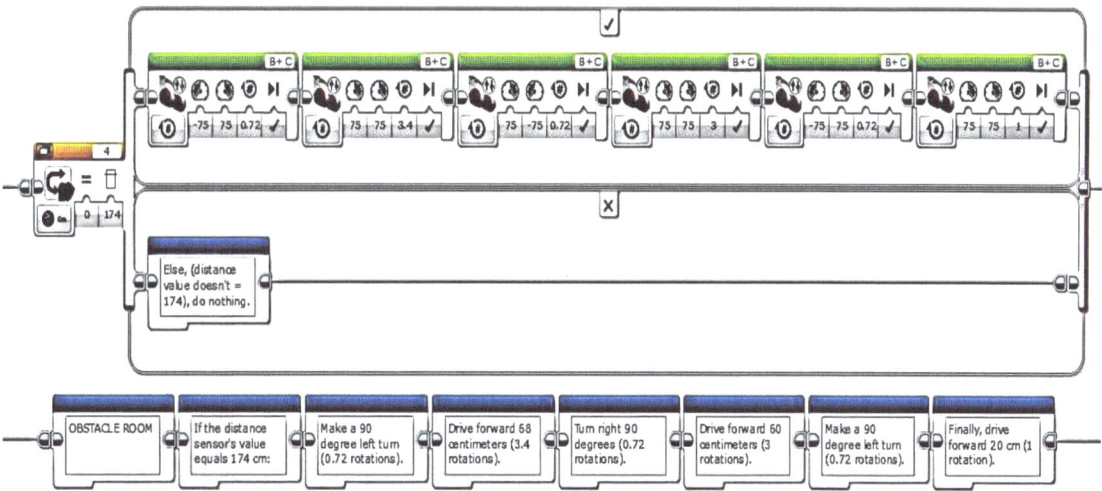

(Diagram is informational only and may not work for your particular robot program).

In the room where there are obstacles, the sensor will turn left, or right depending on the 90-degree angle of rotation, if it senses objects in the room. The robot is also programmed to move forward at one rotation, which would mean one full tire rotation based on the size of the tires. The robot that you build in school will have tires that are small, so imagine that if this was a larger vehicle used for real searches and rescues. The values given in these parameters would be much higher.

The blocks in the flow chart allow for several things to happen. We can apply some mathematical computations to this to perform the programs functions.

Drive Forward- Let's Do the Math

The first step to calculate the measurement of one rotation would be to use a measuring stick and measure one rotation. That will give you the distance the robot will travel based on one tire rotation. Then you can configure the left and right angles and program your robot to turn 90 degrees.

The circumference of the wheel is calculated using the formula: C = 3.14 * d

The EV 3 robot tire is 56mm (there are several sizes but we will us this tire size) the distance we want the tire to move is one rotation. Wheel Converter calculator can be found online. http://ev3lessons.com/resources/wheelconverter/

1 Rotation = 56 mm and 5.6 cm (inches 2.2047)

Input the distance you would like the robot to move in either inches or centimeters

Total distance to travel (in cm) 10.16

Circumference of your wheel (in cm) is (π * diameter): 17.59291886010284

The number of rotations your robot needs to move is (distance to travel ÷ circumference): 0.577505079219163

The number of degrees your robot needs to move is: (rotations × 360): 207.9018285188987

Find the Survivor

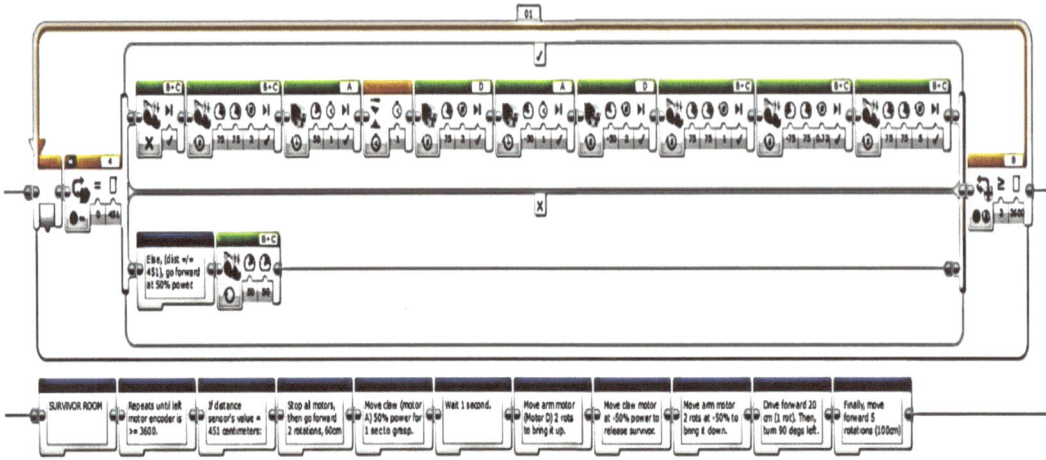

The robot program has a repeat block and will move left until the sensor that is attached to a rotating object (wheel axle). Also called a motor encoder (A measuring device for motion control) measures velocity, the angle of a rotating sensor, acceleration, and displacement. The distance sensor will eventually sense the survivor and move forward, activate the claw motor, pick up the survivor, and if possible in a real world scenario, bring back the survivor and make some sounds to alert rescue workers that a person has been found.

Fire in the Room

This program example is coded to find the fire; it uses a line tracking block with the LED sensor that will follow a line until it finds the fire. The robot rolls over the (fire) and extinguishes the fire.

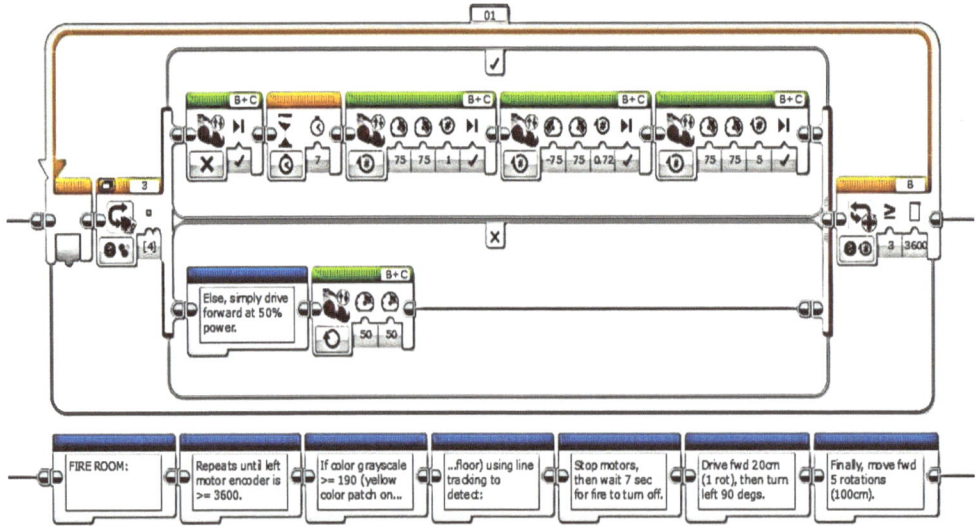

These types of robots that can detect fire, temperatures, and put out fires, are in existence today. Some are built like snakes that navigate to the fire carrying a fire extinguisher to put out the fire. One such robot is named **Anna Konnda**,[7] A water-powered hydraulic robot snake. It is driven by twenty custom-built water hydraulic cylinders. The snake measures 3 meters long and weighs 75 kilograms. The control of this robot is realized using numerous microprocessors that control the joints. The main controller can be connected to a PC via a Bluetooth connection.

LED Room- All Clear?

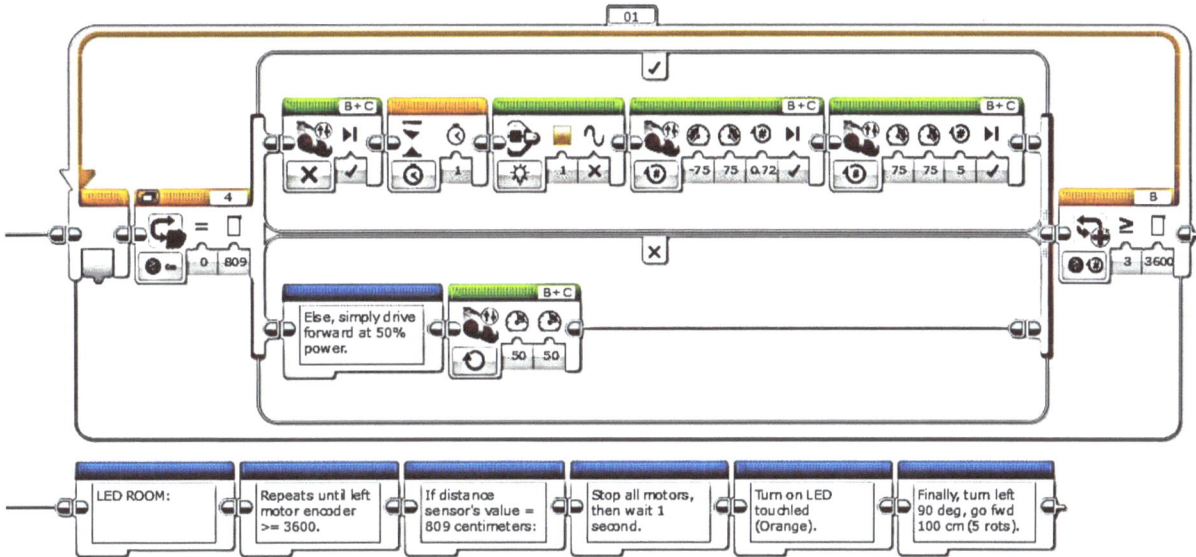

The LED Room

Suppose that the search and rescue team is working in a large area. Time is of the essence to find the survivors. The program is coded to let the rescue team know that a room may be clear. An orange light will come on if there are no obstacles in the room. The robot is going to navigate around the room and come back out, ready to explore another room. "Orange LED" light sensor is programmed to alert the rescue team all is clear in this room. The EV3 codes follow parameters within each block. The B & C servo motors repeat until left motor encoder is equal to or greater than 3600. If the distance sensor value = 809 centimeters, all the motors will stop and wait one second. The LED (orange will turn on). Finally, the robot will turn left 90 degrees and go forward 100 cm (5 rotations) exiting the room since it is all clear.

Similar features are designed in the EV3 LED sensor. The robot can be programmed with color and effects. Three colors, green, red and orange, can be programmed with "effects" of normal, pulse, or flash, depending on what command you want the LED light show.

---

[7] Clarke, Meadow. "Anna Konnda / A Firefighting Robot." Prezi.com. Prezi, 04 Nov. 2015. Web. 03 June 2016. https://prezi.com/hbl4zig3ywae/anna-konnda-a-firefighting-robot/

Segway

For this design, imagine you need to build a machine that will help transport workers around the 2020 Super Bowl Game. The machine must be able to carry persons up to 250 lbs. and have a battery power to last 10 hours. It is an upright self-balancing mobile machine.

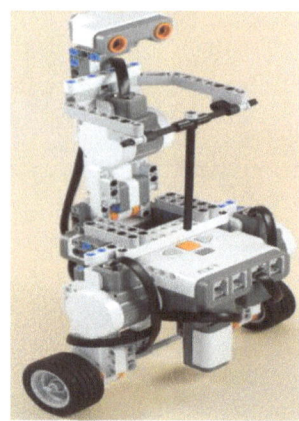

("NXT Segway with Rider" Permission to use Dave Parker, NXTPrograms.com 2016).

The Segway Program

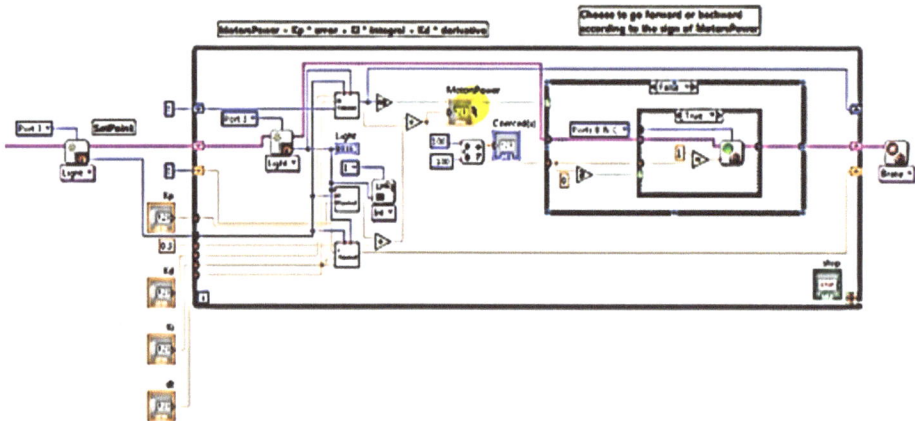

This is schematics from Lab View.

How a Segway functions. The variables determine how fast, forward, or backward your robot will move. The light sensor does a balancing act. It leans forward and back to tell how fast to move based on a range of (Positive) 100 to (Negative) 100. There is a countdown feature to allow for the positive upright balance of the light reading. The set point has reference values and error codes that help to adjust the motor power. If the reading is a positive number, the motor moves forward. If the reading is negative, the motor moves backward. It has a main loop with repeats that are all synchronized to keep the person balanced. The picture shown of the NXT Segway uses a color sensor to balance the Segway.

Segway is one such mobile robot that works this way. It was developed by Dean Kamen in 1999. The main uses for SEGWAY, other than personal use, are for law enforcement and for workers needing to travel great distances during their work day. In August 2008, Beijing Summer Olympics deployed more than 100 SEGWAYS to be used by the Olympic delegates. It was a valuable robot that saved many hours from walking and lost physical energy helping the people to do a job faster with less effort.

The Designer who solved a specific problem "Baxter."

Baxter is the creation and design of Rodney Brooks, MIT Engineer. He also created the famous iRobot Roomba. Baxter has sensors and actuators which mimic human reaction. He plugs into a standard power outlet. Baxter is an affordable, flexible and "safe" robot. What can he do? Line loading, machining tending, package and handle materials, and kitting are just a few to mention. Humans have difficulty with factory work and often quit, or get injured in most tedious factory assembly work. Baxter was designed to free up that job by being able to do these monotonous jobs. He is manually trainable by staff and needs no programming. He is meeting the needs of integrated customizable robots that work with other automation used in the factories.

Used with Permission: Photo by Steve Jurveston 8000143255_e02292c05d_z 07/19.2012 Flickr.com

The Future of Robotics and Real World Applications

Technology advances seem continuously meshed with robotics. The future of robotics appears to be driven by needs of industry, entertainment, military, medical and science. The designs are advancing from industrial needs to robots that are more revolutionary, ones that will work alongside humans. Also, there are tiny robots that are as small as an insect, for robotic "swarms," teams of robots that work collaboratively, fast, cheap and efficiently. We may want robots to look and think like us, but can they? Robots will always need the mastermind of humans, until one day, we program a robot that can completely be programmed by another robot. However, someone will need to be around to fix it when it gets out of sync! As we know it today, man and machine will need to coexist. The designer featured in this chapter Rodney Brooks is the developer of the research work, "Fast Cheap and Out of Control" Thesis. In 1989, Rodney was one of the first to consider the use of tiny robots "swarms."

Complex systems and complex missions take years of planning and force launches to become incredibly expensive. The longer the plan and the more expensive the mission, the more catastrophic if it fails. The solution has always been to plan better, add redundancy, test thoroughly and use high-quality components. Based on our experience in building ground-based mobile robots (legged and wheeled) we argue here for cheap, fast missions using scores of mass produced simple autonomous robots that are small by today's standards (1 to 2 Kg). We argue that the time between mission conception and implementation can be radically reduced, that launch mass can be slashed, that totally autonomous robots can be more reliable than ground controlled robots, and that scores of robots can change the tradeoff between reliability of individual components and overall mission success. Lastly, we suggest that within a few years, it will be possible at a modest cost to invade a planet with millions of tiny robots.[8]

References

Titech to Expand Sensor Based Sorting Range with … (07/18/2016)). Retrieved from https://waste-management-world.com/a/titech-to-expand-sensor-based-sorting-range

Meet Amazon's New Robot Army Shipping Your Product." YouTube. N.p., 14 Dec. 2014. Web. 20 May 2016. https://www.youtube.com/watch?v=g6DIFpaoI6A

VEX IQ Robotics Camp Handbook - manualzz.com. (06/28/2016). Retrieved from http://manualzz.com/doc/7474775/vex-iq-robotics-camp-handbook

Titech to Expand Sensor Based Sorting Range with (06/27/2016). Retrieved from https://waste-management-world.com/a/titech-to-expand-sensor-based-sorting-range

Flowcharts." Springer Reference (2006): 1-3. 4-h CCS. National Robotics Engineering Center, 2006. Web. 22 May 2016. http://www.education.rec.ri.cmu.edu/robots/4H/roboticsandyouonline/explorer_content/explorer_7/flowcharts.pdf

Clarke, Meadow. "Anna Konnda / A Firefighting Robot." Prezi.com. Prezi, 04 Nov. 2015. Web. 03 June 2016. https://prezi.com/hbl4zig3ywae/anna-konnda-a-firefighting-robot/

Brooks, Rodney. "Fast, Cheap, and Out of Control: A Robot Invasion Thesis." CiteSeerX. Journal of The British Interplanetary Society, 1989. Web. 2016. <http://citeseerx.ist.psu.edu/viewdoc/summary?doi=10.1.1.132.4580>. Vol.42, pp 478-485

---

[8] Brooks, Rodney. "Fast, Cheap, and Out of Control: A Robot Invasion Thesis." CiteSeerX. Journal of The British Interplanetary Society, 1989. Web. 2016. <http://citeseerx.ist.psu.edu/viewdoc/summary?doi=10.1.1.132.4580>. Vol.42, pp 478-485

CHAPTER 6

Mathematic Applications with Robots

Mathematics is pretty important in Robotics. Think about all the tasks a robot can perform. Mathematics make the tasks happen. Math is found in all aspects of robotic engineering. Some basic examples of robotic engineering are used manufacturing. Math is needed to calculate motor power or torque required to lift an object. **Variables** are programmed to calculate the weight that is lifted in an object, the ratio of the gears to speed up or speed down the machine, and the length of the robotic arm Degrees of Freedom. The speed at which the object needs to be lifted can also vary. Extra weight added to the robot for strength decreases the amount of the load the robot can lift. Precise math calculations are used to minimize the material used while maximizing the available lifting power of a robot.

Engineers use math for several reasons: First, the Laws of Nature (E.G., Maxwell's Equations for Electromagnetics, Kirchhoff's rules for Circuit Analysis) are mathematical expressions needed to make a robot mobile. Mathematics is the language of physical science and engineering. Second, mathematics is more than a tool for solving problems; a mathematic course can help a student develop intellectual maturity. You can stretch your knowledge and understanding of the physical world by understanding math principles. Last, computers do not make traditional mathematical analysis obsolete. The reason for this is that computer programs contain mathematical relations; understanding these relationships is still necessary.

Preparation for a career in the field of robotic engineering will require coursework in College Algebra, Trigonometry, Calculus I and II, Geometry, Linear Algebra, Differential Equations, and Statistics. If you decide on a career as a Robotic Engineer, your studies will include designs and mathematics, debugging programs, servicing and calibrating robots courses, integration of robots with mechanical devices, and supervision of others. Knowledge of mathematics, computer programming, computer-aided design, and drafting are intertwined as you develop skills and experience. If you are in high school, take the advanced Math and Science Courses. It will help to prepare you for a career in Robotics.[1] If you can, participate in a robotic club, one that has competitions that will give you practical experience working with robots, as well as experience to add to your resume' when applying for college admission. Math is an important skill in the advanced domain of robotics.

Examples of Applied Mathematics

Wheel Size Diameter and Linear Distance Traveled

("SevenStuds – Custom 3D Printed Wheels + LEGO Creations", 2016).

Finding the correct wheels to fit parameters and perform a task will help you to overcome navigation and technical issues. Distance traveled is equal to the length of the wheel circumference as it rolls along the floor and the wheel turns. A ruler is necessary to complete the measurement. If you calculate the distance traveled based on the rotations of an axle and the diameter of a wheel attached to the axle, you should figure out that as wheel size increases, distance traveled increased proportionally.

Demonstration

If the wheel completes 4 5/16th revolutions, you can calculate how far it will go.

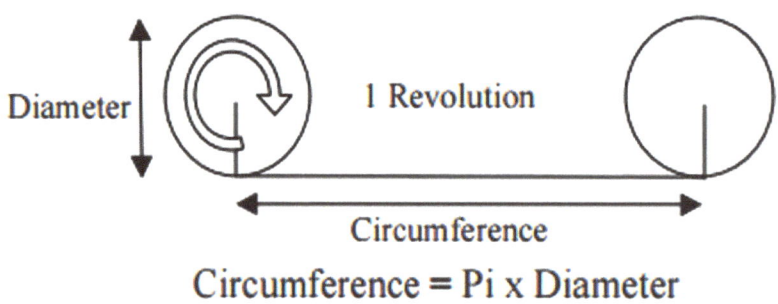

---

[1] How to Become a Robotics Engineer in 5 Steps. (May 29,2016). Retrieved from
http://learn.org/articles/Robotics_Engineering_Become_a_Robotics_Engineer_in_5_S

The circumference of a wheel is equal to its diameter multiplied by Pi (about 3.14).[2]

	Diameter		PI		Circumference
1	1.9375"	X	3.14	=	6.08"
	Circumference		Revolutions		Distance
2	6.08"	X	4.3125	=	26.22"

Circle geometry (Diameter, Circumference) C = π*d

Convert the fractions to decimals. The diameter conversion is 1 15/16th" to 1.9375."

Convert the rotations to decimals. The rotations conversion will be 4 5/16th" to 4.3125."

The circumference is equal to π x D or: 3.14 x 1.9375 = 6.08"

4.3125 revolutions: 6.08" x 4.3125 = 26.22" (total distance traveled)

Measurement Distance - Centimeters

To covert between the unit systems of centimeters and inches apply the math and calculate.

1 rotation = 360 degrees = X cm = Y in. The values of X and Y are calculated based on the wheel size.[3]

Suppose a robot (A) with 2.5-inch wheels and a robot (B) with 3.5-inch wheels (both) drive forward for one minute. They have the same RPM (revolutions per minute). Robot (B) will go farther than robot (A). Robot (B) will have a greater linear velocity. Both (A) and (B) will have the same angular velocity.

Robot (B) is driving with an angular velocity of 20 radians/sec. Calculate Robot (B) linear velocity.

Linear Velocity v = xt

[2] content.teachengineering.org. (May 31,2016.). Retrieved from http://content.teachengineering.org/conmtent/nyu_/activities/n

[3] Diameter Distance Traveled TEACH (May 28th, 2016). http://www.education.rec.ri.cmu.edu/previews/rcx_products/robotics_educator_workbook/content/mech/pages/Diameter_Distance_TraveledTEACH.pdf

Angular Velocity

Angular Velocity is a measure of how quickly an object moves through an angle. It is the change in angle of a moving object (measured in radians), divided by time. Angular velocity has a magnitude (a value) and a direction.

Angular velocity = (final angle) - (initial angle) / time = change in position/time

$\omega = (\theta_f - \theta_i) / t$

ω = angular velocity

θ_f = the final angle

θ_i = the initial angle

t = time

$\Delta\theta$ = short form for 'the change in angle'[4]

$$v = wr \qquad w = \frac{20\text{rad} \quad r = 3.5\text{in}}{\text{sec}} 2$$

$$v = \frac{20\text{rad}}{1\text{sec}} \times 1.75\text{inc} = \frac{35\text{in}}{\text{sec}}$$

Robot (A) is driving with an angular velocity of 20 radians/sec. Calculate Robot (A)'s linear velocity.

$$v = wr \qquad w = \frac{20\text{rad} \quad r = 2.5\text{in}}{\text{sec}^2}$$

What if you want both robots to end up at the same location? Given their different wheel sizes, a good way to figure out the final result would be to have Robot (B) drive for a shorter time than Robot (A). Robot (B) will get there first, and both robots will end up at the same location. Another way may be to have Robot (B) drive with fewer RPM's (or a smaller angular velocity) than Robot (A), then they arrive at the same location at the same time.

Practical Application in the Physical World- Angular Velocity

On road and off road vehicles have different size tires: Sports cars, SUV's trucks tractors, motorcycles, and three-wheelers. These sizes vary for reasons such as attaining higher speeds, depending on the performance characteristics required for the type of vehicle.

---

[4] Angular Velocity Formula. (2016). Softschools.com. Retrieved 30 December 2016, from
http://www.softschools.com/formulas/physics/angular_velocity_formula/21/

Velocity (feet per second) and angular velocity are directly related. W is (in radians per second- not RPM) of the motor and radius of the wheel. R is calculated in meters or feet. So when you calculate V, it is in meters per second.

W = v/R Therefore the velocity of the robot is v = w * R

Most automotive motor manufacturers provide the no load RPM.

To convert RPM to radians per second, you need this equation:

1 rev/min = 2 * pi/60 rad/sec

Equations For Robot Navigations

Calculations are necessary to make decisions for the robot to move. A few equations to use for speed and direction can be figured and setup in the software parameters that you are using to program your robot, and change the parameters to suit the tasks. The method of programming a robotic turn can affect the turn results.

Motor Speed = rotations / time

Point Turn = degrees (1/2 axle length) / wheel radius

Swing Turn = degrees (axle length) / wheel radius

Forward/Back rotations = distance / 2 * 3.14 (wheel radius)

Center of Gravity

A robot can be built to walk. Walking robots offer an opportunity to experiment with the center of gravity. The **Center of Gravity** (COG) in a robot is the "center position" of all the weight of the robot. The COG uses both position and weight; the heavier the objects affect the gravity balance of the robots more than lighter robots. If a robot has object manipulators (arms or claws), that also changes the how you would determine where the center of gravity is going to be.

"Walking Dinosaur" Vex IQ® Robotics (Flicker.com 6-1-2016).

Calculation Suggestions

Calculate the minimum distance required for the robot to make a turn. Use a different motor for each side and increase speed on one side. Configure opposite directions for each side, i.e., forward and backward. Changing the settings of the degrees, time, rotation, and power levels are optimizations that give your robot the ability to center the gravity of the robot.

You can figure how long a step is by considering how a pedometer works. It calculates how many steps you have taken and calculates how far you have walked. Try it yourself. Measure a distance and take several strides to complete that distance. Repeat several times and make the calculation to find the average step length. Next, Program your robot to move a specific number of rotations at least 12 and as with your walking test, the number you choose will be based on the space you have (environment) for your robot. Make sure you start the robot at the same location each time, run the program and measure the distance walked six times. Calculate to find the average robot step length.

Distance 1 +......... Distance 6 (all runs) = Sum of Distances (Total Distance)

Total Distance / Number of Tests Run = Average Distance in a Single Run

Average Distance in a Single Run / Number of Steps in Each Run = Average Length of Step

Does the shape and length of the leg of your robot affect the center of gravity? The model build may need to be modified to keep your robot upright. Do some research on artificial legs. How do they move? Are the different types of mechanisms found in a prosthetic leg similar to a robot?

Speed and Acceleration

Your robot can travel in a straight line. The speed the robot travels is dependent upon the rotational speed of the servomotors, the brick brain power that has been figured into the code, and any gear ratios that you have put together with the diameters of the tires. How fast the drive wheels are spinning their angular velocity can be defined by engine speed and gear ratio. Tire sizes make some differences.

Tires	Small	Medium	Large
	Solid 24mm x 7mm	Solid 30mm x 10.7mm	Solid 43mm x 10.7mm
	Balloon 30.4 mm x 14	Balloon 49.6 mm x 28 mm	Balloon 81.6mm x 15mm

Pulley Wheel 30 mm x 4 mm

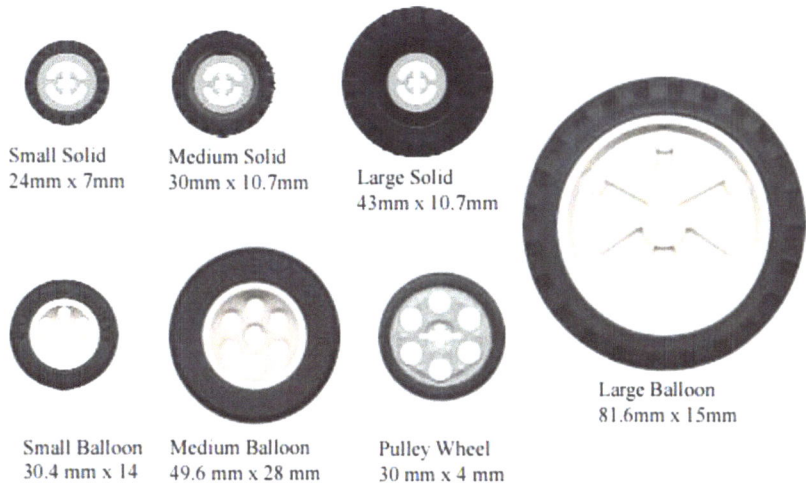

Small Solid
24mm x 7mm

Medium Solid
30mm x 10.7mm

Large Solid
43mm x 10.7mm

Small Balloon
30.4 mm x 14

Medium Balloon
49.6 mm x 28 mm

Pulley Wheel
30 mm x 4 mm

Large Balloon
81.6mm x 15mm

These are the tire sizes that work with most robots.

Use the equation for the circumference of a circle to convert angular velocity to linear velocity.

Demonstration

A typical robot running on a flat terrain will need acceleration to be half of your max velocity. If the robot velocity is 3ft/s, and you need the acceleration to be close to 1.5 ft./s^2, it would take 2 seconds (3/1.5 = 2) to reach the maximum speed.

Remember that: **Force = Mass * Acceleration**

Another factor to consider when choosing acceleration is that if the robot is going through rough terrain, or up inclines, the robot needs a higher acceleration due to countering gravity. If the robot were to go straight up a wall, it would need an extra 9.81 m/s^2 (32 ft./s^2) of acceleration to counteract the gravity. A 20-degree incline would require 11 ft./s^2.

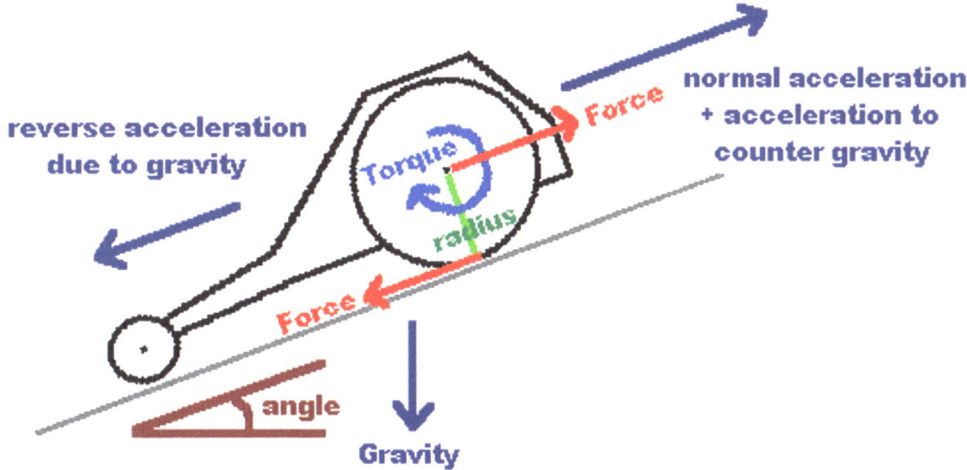

Torque: Society of Robots Can use with Permission: attribute to Society of Robots, 2016).

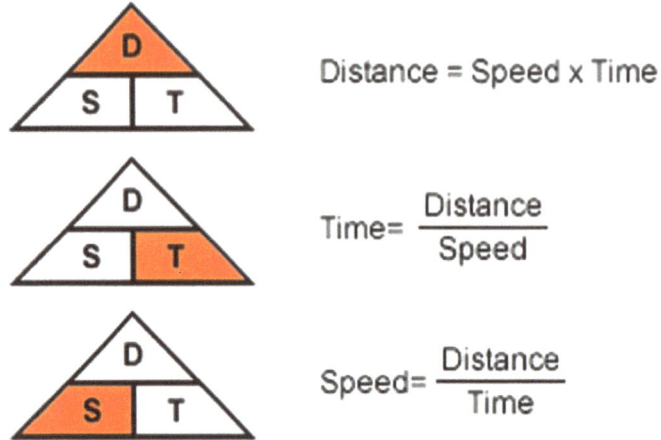

To calculate additional acceleration for a specific incline use the formula

Acceleration for inclines = 32 ft./s^2 * sin((angle_of_incline * pi) / 180).

The acceleration must be added to what you already calculated to move on a flat terrain.

Motor acceleration and torque are not constants. Motor acceleration will decrease as motor rotational velocity increases.[5]

Speed equation (d = rt) the triangle is a way to remember how to divide or multiply to find the distance, time and speed.

Analyzing Data Trials

You can apply some mathematical computations to analyze data that is collected while doing an experiment. First, think of a hypothesis for before beginning the experiment. In this case, the hypothesis will be, "Will the robot move the same distance in all three trials." The actual distance traveled in three separate trials is based on measurements taken from the wheel and distance traveled. There are three trials within this condition, and we will figure out if the robot will go the same distance in all three trials.

Factors such as unpredictable wheel slippage, motor power variation, and human error (aiming the robot) are considered "random experimental errors" and will keep your robot from having the same measurement.

Demonstration

Calculate the average distance that the robot will go with your wheel measurement

$$\text{Average (Mean) Distance} = \frac{\text{distance 1 distance} + 2 \text{ distance} + 3 \text{ distance}}{3 \text{ Trials}}$$

---

[5] How to Build Robot Tutorials - Society of Robots. (May 24, 2016)). Retrieved from
http://www.societyofrobots.com/mechanics_dynamics.shtml

This is the average of the values of the three trials for this task. Averaging the data helps to reduce the effects of the random experimental error. This will help not only to reduce error but to get a better calculation of the "true" distance. Using this method will work because (example) a slightly-too-high value will cancel out a slightly-too-low value when they are averaged, leaving a more "correct" value in general.

$$\frac{\text{Theoretical Measurement} - \text{Actual Measurement}}{\text{Theoretical Measurement}} \times 100\%$$

Goodness Fit- % error calculation will use the theoretical measure and is a quick estimate to measure how well the theoretical measurement measures the actual data. When working with a hypothesis, this is a valuable way to prove or disprove how valid your hypothesis may be.

Based on discovered or proposed linear/optional relationships, speed and distance the robot will travel when changing wheel sizes/RPM and the number of rotations the axle can make can be calculated based on known measurements and time.[6]

You can use some math formulas to understand and figure out how fast a robot moves across a floor based on how fast the wheel spins. Each time the wheel makes a full revolution (rotation), it will roll forward at the same distance equal to its circumference. Calculate the circumference of the wheel to get the exact speed the robot is moving per one revolution.

Demonstration

To do this example: the wheel diameter is 4 inches and the wheel spins at 50 RPM. How fast is the wheel rolling in inches per second?

Circumference = Diameter x Pi

Circumference = 4 Inches x 3.14

Circumference – 12.56 inches

Now determine the terrain speed based on the wheel (Rotations Per Minute) RPM. A great tool to remember is Dimensional Analysis, or some refer to it as Factor Conversion.

$$\frac{12.56 \text{ inches}}{1 \text{ revolution}} \times \frac{50 \text{ revolutions}}{1 \text{ minute}} = \frac{628 \text{ inches}}{1 \text{ minute}}$$

$$\frac{628 \text{ inces}}{1 \text{ minute}} \times \frac{1 \text{ minute}}{60 \text{ seconds}} = \frac{10.467 \text{ inches}}{1 \text{ second}}$$

---

[6] content.teachengineering.org. (May 31,2016.). Retrieved from http://content.teachengineering.org/conmtent/nyu_/activities/n

The next step may be to calculate the gear reduction based on your robot motor speed of approximately 100 RPM. If your wheel is 5 inches in diameter, you want the robot to travel at 4 feet per second, and the motor spins at 100 RPM, calculate the circumference of the wheel and convert the target speed to inches per second.

Circumference = 5 in x 3.14 = 15.7 inches

$$\frac{\textbf{4 ft.}}{\textbf{1 second}} = \frac{\textbf{48 in}}{\textbf{1 second}}$$

If you move 48 inches in one second, and each revolution is 15.7 inches; we can calculate the revolutions per second the wheel will turn to achieve 4 feet per second.

$$\frac{48 \text{ inches}}{1 \text{ second}} \times \frac{1 \text{ revolution}}{15.7 \text{ inches}} = \frac{3.057 \text{ revolutions}}{1 \text{ second}}$$

$$\frac{3.057 \text{ revolutions}}{1 \text{ second}} \times \frac{60 \text{ seconds}}{1 \text{ minute}} = \frac{183.42 \text{ revolutions}}{1 \text{ minute}}$$

Basic programming - Conditional Statements, Calculating Thresholds, Solving Equations

Robot lessons involve algebra. It includes thresholds and conditional statements (inequalities), programming sensors, measuring turns (equalities, solving equations), gear and speed ratios of direct and indirect proportionality. In ordinary algebra, variables take on real value. Behaviors of the variables interact with ordinary operations of arithmetic, namely, addition, subtraction, multiplication, and division. In Boolean algebra, the variables have only two values, true or false. The operations become symbolic logic rather than arithmetic and are called conjunction (*and*), disjunction (*or*), and negation (*not*).

Truth Statements

To program a robot, you become the decision maker. You program the robot to make clear choices based on the conditions or the circumstances. These are called Boolean Statements, not questions. Rather than asking "is the light red" and get a yes or no, a Boolean Statement will give a value of true= it is red or false= it is not red.

Conditions

The brick brain is programmed to make a decision based on the codes to run, and the math applied to the power, speed, and tire diameter. Everything works precisely and integrates with the conditions of the environment. The decisions may be to repeat (loop block) and if-else conditional statements. The conditions in robotics are Boolean statements.[7]

---

[7] Boolean Logic - RobotC. (May 31, 2016.). Retrieved from
http://cdn.robotc.net/pdfs/nxt/reference/hp_boolean_algebra

Demonstration

Condition	Ask....	Boolean Statement
Sensor Value (Light Sensor) ≥ 57	Is the value of the light sensor greater than 57?	True, when the current value is more than 57 False, if the current value is not more than 57

Robotic Symbols and their Meaning

Comparisons can be used to compare the values of the condition. The light sensor above has a value of ≥57. It is a specific operation that must be performed. It will also produce a true or false statement result. The software program RobotC symbols chart is an example of the symbols and their meanings.

Robot C Symbol	Value	Comparison Example	Outcome
= =	Equal to	50 = = 50 40 = = 100	True False
!=	Is not equal to	50 != 50 90 != 40	True True
≤	Is less than	20 ≤ 85 20 ≤ 10	True False
≤ =	Is less than or equal to	50 ≤ = 0 50 ≤ = 100	False True
≥	Is greater than	60 ≥ 20 60 ≥ 100	True False
≥ =	greater than or equal to	40 ≥ = 100 65 ≥ = 40	False True

Measuring Distance with the Ultrasonic Sensor

We can measure distance and time using a robotic sensor. In this example, we will use the NXT Ultrasonic sensor. First, we need to obtain a few measurements. Three measurement readings of distance sensor sound at room temperature (a pseudo reading will be applied to this example). Obtain an average of the three readings.

Demonstration

Table 1

Distance to Object 1	Distance to Object 2	Distance to Object 3	Distance to Object 4
68 cm	69 cm	68 cm	68.3 cm

By changing the centimeters into meters, we can calculate the distance. The other information that is needed is the time it takes for the sound wave to travel (round-trip) from the sensor to the object and back again. Convert the round-trip time of a sound wave from seconds to microseconds.

1 second = 1,000000 microseconds or 1 microsecond = 1-6 seconds

Table 2

Distance to Object Average (meters)	Speed of Sound(m/s)	Time to the Object(s)	Round-trip Time(s)	Round-trip Time Microseconds
0.683	343.6	0.002	0.004	4000

Time to the object = Distance between ÷ Speed of sound = 0.683(m) ÷ 343.6 (m/s) = 0.002s

Time round-trip (s) = 2 *

Time between = 2 * 0.002 (s) = 0.004 (s)

Time round-trip (microseconds) = 4000[8]

---

[8] Measuring Distance with Sound Waves Activity – Distance … (May 27, 2016). Retrieved from http://content.teachengineering.org/content/nyu_/activities/n

Frequency

The frequency of a wave is understood as the number of cycles a wave completes in one second. We calculate the frequency of the wave of 10Hz, completes 10 full cycles in one second. The wave will complete 1 cycle in 0.1 seconds or 100,000 microseconds.

A wave completes 10 cycles in 1 second, hence 1 cycle is completed after x number of seconds.

Set up a proportion **10(cycles)** = 1 and **(cycle)** you can solve for x, and convert the
          1(sec)   **(sec)**
result into microseconds. It takes 100,000 microseconds for a wave to complete 1 cycle, then after 4,000,000 microseconds, the wave complete

A LEGO Ultrasonic Sensor wave frequency is 40 000 Hz. 40,000.

How many cycles can it take? The ultrasonic sensor wave travels one cycle and is based on Hertz 25 microseconds.

Note: Methods of solving this question may vary. A wave completes 40000 cycles in 1 second, after x number of seconds one cycle is completed. Set up a proportion and solve for x.

$$\frac{40000 \, (cycles)}{1(s)} = \frac{1(cycle)}{x(s)}$$

$$X = \frac{1(s) * 1(cycle)}{40000 \, (cycles)}$$

How many cycles does the Ultrasonic sensor wave go through, traveling from a sensor to the object and back? We use the calculated round-trip time in Table 2.

160(cycles) It takes 4000 microseconds for a wave to travel from the ultrasonic sensor to an object and back. It takes 25 microseconds for the Ultrasonic wave to make one cycle. Therefore,

$$40000 \, (microseconds) * \frac{1 \, (cycle)}{25 \, (microseconds)} = 160 \, (cycles)$$

Gear Ratios

Gears are the force that is generated at the edge of the gear, and the force equals the product of the radius of the gear along with its torque in a line tangential to its circumference. The amount of rotational force or pushes you have on the gear is called torque. The faster the gear turns, the lower the torque. The slower the gear turns, the higher the torque.

Need More Torque?

Your robot may not move fast enough to complete the task you have programmed it to do. You might be trying to go up a hill or push something heavy. *If you lift the robot in the air and run the program, the wheels turn in the air but not on the ground you, need more torque.*

When the follower turns faster than the driving gear it is called gearing up. The driver gear may have more teeth than the follower. Gearing up is great for getting your robot to go fast when it does not need a lot of torque. Gear down so that the follower turns slower than the driving gear. The driving gear has fewer teeth than the follower gear. Gearing down is when you need your robot to have a lot of torque, and speed is not important.

Remember those simple machine lessons in elementary school? The driver gear turns clockwise, and the follower gear turns counter-clockwise. You can find a Gear Ration Calculator on this site ("LEGO Gear Ratio Calculator", 2016) an online tool for gearing up and gearing down.

Force = radius x torque.

When gears are combined with different radii, the amount of force/torque the mechanism generates can be controlled. The torque and radii relationship is found as follows:

Gear 1 has a radius r1 and turns with torque t1. It is generating a force of t1/r1 perpendicular to its circumference. It meshes with Gear 2, and with r2, which generates t2/r2, then t1/r1 = t2/r2. The torque gets generated by Gear 2; the answer is t2 = t1 r2/r1

The torque is generated at the output gear. It is also proportional to the torque that is the input gear and the ratio of the two gear's radii. When r2 > r1, we get a large number, if r1 > r2, we get a small number.

Demonstration

Basic Gear Ratio Theory

$T_i = F \times R_i$
$T_o = F \times R_o$

VELOCITY
$V_o = V_i \times \dfrac{R_i}{R_o}$

TORQUE
$T_o = T_i \times \dfrac{R_o}{R_i} = T_i \times \dfrac{V_i}{V_o}$

When output gears are larger than the input gears, torque increases. When an output gear is smaller than the input gear, torque decreases. when gears are combined to have a change in torque, the corresponding speed changes.

Gear Output/Input

The circumference of the gear (C = 2 * pi * r) is used to measure speed. If the circumference of Gear 1 is twice the size of Gear 2, Gear 2 must turn two times for each full rotation of Gear 1. If the output gear is larger than the input gear, the speed decreases. If the output gear is smaller than the input gear, the speed increases. When a small gear drives a large one, torque is increased, and speed is decreased. Thus, when a larger gear drives a smaller gear torque is decreased, and speed is increased. DC motors use gears that are fast with low torque. The tradeoff is extra speed for additional torque.

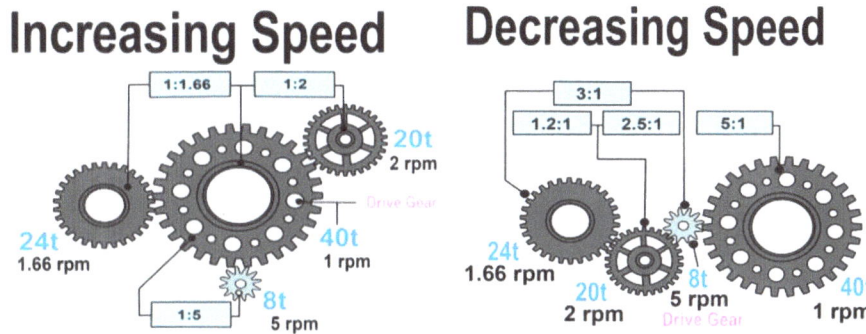

Gears are combined using their teeth. The gear teeth are specially designed so that they mesh properly. The number of teeth is not random. They are designed to be a ratio between the rate of the last and first gear teeth as it cycles through one cycle. The gear turns need proper reduction. Backlash occurs if there is looseness between the meshing gears. The mechanisms of the *gears need to h*ave the ability to move back and forth within the teeth without turning the whole gear and being too loose.

To reduce the backlash, tight meshing between the gear teeth can increase *friction*. Manufacturing and designing gears can be complicated.

To apply power to a small gear, (one 8-teeth gear) meshed with a large gear (3 * 8 = 24 tooth gear). We can achieve a "three to one gear reduction (3:1)." As a result, the large gear slows down by 3 and the torque is tripled. A "ganged" (gear is organized in series to multiply their effect). Ganged gears such as those found in electrical devices and machines have gears arranged to increase or decrease torque.

Another example is 2 3:1 gears in a ganged series that results in a 9:1 reduction. The gears will be optimally arranged so that three 3:1 gears in a series can produce a 21:1 reduction. A multiplying reduction is the best method mechanism for making DC motors useful.[9]

---

[9] content.teachengineering.org. (May 31,2016.). Retrieved from
http://content.teachengineering.org/content/nyu_/activities/n

LEGO Gear Ratio Table

Follower Gear on Top Driver Gears along the Side of the Table

Number of Teeth	1	8	12	14	16	20	24	28	36	40	56
1	1:1	1:8	1:12	1:14	1:16	1:20	1:24	1:28	1:36	1:40	1:56
8	8:1	1:1	1:1.5	1:1.75	1:2	1:2.5	1:3	1:3.5	1:4.5	1:5	1:7
12	12:1	1.5:1	1:1	1:1.16	1:1.33	1:1.67	1:2	1:2.33	1:3	1:33.3	1:4.67
14	14:1	1.75:1	1.16:1	1:1	1:1.14	1:1.42	1:1.71	1:2	1:2.57	1:2.86	1:4
16	16:1	2:1	1.33:1	1.14:1	1:1	1:1.25	1:1.5	1:1.75	1:2.25	1:2.5	1:3.5
20	20:1	2.5:1	1.67:1	1.42:1	1.25:1	1:1	1:1.2	1:1.4	1:1.8	1:2	1:2.8
24	24:1	3:1	2:1	1.71:1	1.5:1	1.2:1	1:1	1:1.17	1:1.55	1:1.67	1:2.33
28	24:1	3.5:1	2.33:1	2:1	1.75:1	1.4:1	1.17:1	1:1	1:1.29	1:1.43	1:2
38	38:1	4.5:1	3:1	2.57:1	2.25:1	1.8:1	1.55:1	1.29:1	1:1	1:1.11	1:1.55
40	40:1	5:1	3.33:1	2.86:1	2.5:1	2:1	1.67:1	1.43:1	1.11:1	1;1	1:1.4
56	56:1	7:1	4.67:1	4:1	3.5:1	2.8:1	2.8:1	2:1	1.55:1	1.4:1	1:1

This chart will help make finding the gear ratios simple. Your answer may be a decimal, fraction or ratio (i.e., X: Y)

Geometry- Navigating Robot Parameters

Geometry can be applied when you need to figure out the Euclidean space (a 2D plane length and width). In an advanced class, you might want to program a robot in a 3D space, by calculation the "reach" of a robot arm along with the length and width parameters of the robot's environment. The (W) is called the workspace. It helps to make the decisions for mobility of your robot. What you want a mobile robot to do can be explained and documented, before writing the code. On the ElectronicTeacher© website, you can find a concise webpage of advanced robotic and electronic information. Below is an *excerpt* from "**tasks and mobility of robots.**" In order to achieve the goals to complete the tasks of navigating physical objects, you could set up an example of a robot navigating an area that has obstacles in its path as given below.

A is a single rigid object represents the **_robot._** The Euclidean space **W** is the **_workspace_** and is represented as R^n n- 2 or 2). $B_1.....B_q$ is fixed rigid objects found throughout **W**. These are the **_obstacles_**. In the geometry of **A**, the geometries and locations of B_i's are accurately known. Assuming that no kinematic constraints limit the motions of **A**, **A** is a **_free-moving object_**. The initial position, orientation and goal position of **A** is given in **W**; the problem is to generate a _path_ **T**. It is specified as a continuous sequence of position and orientations of **A**. It will avoid contact with B_i's. If you give the <u>robot</u> a bunch of objects to avoid, a start state, and a goal state, you can find (calculate) the path the <u>robot</u> will take to reach the goal state.[10]

Schematics

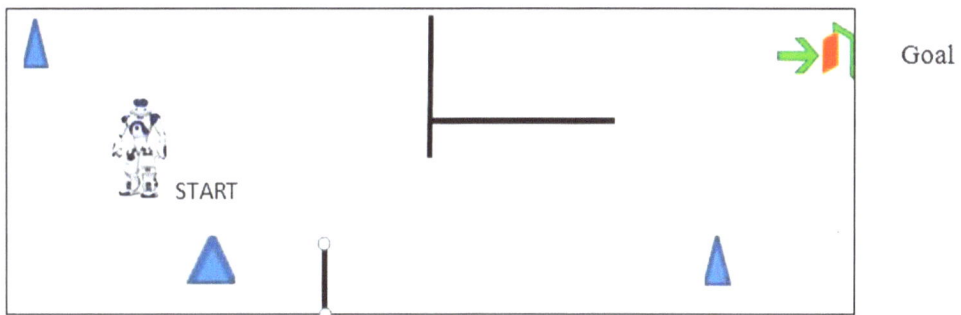

W = Workspace A = Robot (free moving) T = Path B = Objects

In RobotC a program written would follow pseudocode statement, commands and 0 respectively. As the robot begins to navigate through the building, it encounters objects, commands that it is true to move forward, the sensor detects the object at a distance of 19 cm. Both motors will stop. The robot will point 180 degrees and turn 45 degrees clockwise. Once the robot has reset its path, it continues to navigate forward until it encounters another robot.

Startup();
driveForward(Unlimited,100);
Drives forward. Optional parameters include distance and power level. If no distance is specified, use one of the "stop" or "until" functions to stop the robot.
(
{ while(true) –
(
{ if(Sensor Value (Ultrasonic Sensor)>19) **(19 represents the distance the sensor can detect an object in centimeters)**
{ motor[motorC]=100;
motor[motorB]=100; } **both motors run full speed if the commands are true**
)
else stopIfSonarLessThan 19);

---

[10] Retrieved from: http://www.electronicsteacher.com/robotics/robotics-planning.php#Task-Planning

www.ingramcontent.com/pod-product-compliance
Lightning Source LLC
Chambersburg PA
CBHW050852180526
45159CB00007B/2648